Alexander Strehblow

Modulators of Nuclear Localization of the Human Enzyme ADAR1

Alexander Strehblow

Modulators of Nuclear Localization of the Human Enzyme ADAR1

DsRNA-binding domains of ADAR1

Südwestdeutscher Verlag für Hochschulschriften

Impressum/Imprint (nur für Deutschland/ only for Germany)
Bibliografische Information der Deutschen Nationalbibliothek: Die Deutsche Nationalbibliothek verzeichnet diese Publikation in der Deutschen Nationalbibliografie; detaillierte bibliografische Daten sind im Internet über http://dnb.d-nb.de abrufbar.
Alle in diesem Buch genannten Marken und Produktnamen unterliegen warenzeichen-, marken- oder patentrechtlichem Schutz bzw. sind Warenzeichen oder eingetragene Warenzeichen der jeweiligen Inhaber. Die Wiedergabe von Marken, Produktnamen, Gebrauchsnamen, Handelsnamen, Warenbezeichnungen u.s.w. in diesem Werk berechtigt auch ohne besondere Kennzeichnung nicht zu der Annahme, dass solche Namen im Sinne der Warenzeichen- und Markenschutzgesetzgebung als frei zu betrachten wären und daher von jedermann benutzt werden dürften.

Verlag: Südwestdeutscher Verlag für Hochschulschriften Aktiengesellschaft & Co. KG
Dudweiler Landstr. 99, 66123 Saarbrücken, Deutschland
Telefon +49 681 37 20 271-1, Telefax +49 681 37 20 271-0, Email: info@svh-verlag.de
Zugl.: Wien, Universität Wien, Dissertation, 2005

Herstellung in Deutschland:
Schaltungsdienst Lange o.H.G., Berlin
Books on Demand GmbH, Norderstedt
Reha GmbH, Saarbrücken
Amazon Distribution GmbH, Leipzig
ISBN: 978-3-8381-0361-7

Imprint (only for USA, GB)
Bibliographic information published by the Deutsche Nationalbibliothek: The Deutsche Nationalbibliothek lists this publication in the Deutsche Nationalbibliografie; detailed bibliographic data are available in the Internet at http://dnb.d-nb.de.
Any brand names and product names mentioned in this book are subject to trademark, brand or patent protection and are trademarks or registered trademarks of their respective holders. The use of brand names, product names, common names, trade names, product descriptions etc. even without a particular marking in this works is in no way to be construed to mean that such names may be regarded as unrestricted in respect of trademark and brand protection legislation and could thus be used by anyone.

Publisher:
Südwestdeutscher Verlag für Hochschulschriften Aktiengesellschaft & Co. KG
Dudweiler Landstr. 99, 66123 Saarbrücken, Germany
Phone +49 681 37 20 271-1, Fax +49 681 37 20 271-0, Email: info@svh-verlag.de

Copyright © 2009 by the author and Südwestdeutscher Verlag für Hochschulschriften Aktiengesellschaft & Co. KG and licensors
All rights reserved. Saarbrücken 2009

Printed in the U.S.A.
Printed in the U.K. by (see last page)
ISBN: 978-3-8381-0361-7

Table of Contents:

Abstract:	3
Zusammenfassung:	4
1. Introduction:	5
RNA modification:	5
Processing of eukaryotic RNAs:	
RNA Editing:	
Ribonucleotide deamination:	
ADARs (Adenosine deaminases that act on RNA):	10
Discovery of ADARs:	
Substrate recognition and substrates of ADARs:	
Biological importance of ADARs:	
Members of the human ADAR family:	
Functional domains of ADAR:	
Nuclear transport:	20
The Nuclear Pore Complex (NPC):	
The RanGTPase cycle:	
Transport factors, cargos and their localization signals:	
Nuclear transport of hsADAR1:	28
Specific aims of the project:	30
2. Material and methods:	32
Cloning and plasmids:	
Expression and purification of recombinant proteins:	
Tissue culture:	
Exportin-5 RNAi:	
Preparation of cytosolic, nuclear and total cell extracts:	
Immunoprecipitations:	
Gelelectrophoresis and western blotting:	
Import assays:	
Affinity chromatography and mass spectrometry:	

3. Results: 41
Nuclear Import of hsADAR1: 41
Structural analysis of dsRBD3 as an active NLS:
Interaction of import receptors with dsRBD3 *in vitro*:
DsRBD3 can be imported by several karyopherins in import assays:
Transportin-1: a possible candidate:

Modulation of ADAR1's nuclear concentration: 58
Possible functions of the Modulator Of Import (MOI):
Exp5 binds several dsRBDs in a RanGTP- and RNA-dependent manner:
No indications for NES activity of the MOI *in vivo*:
No evidence for export via import assays:
Knock down of Exp5 by RNAi:
Ongoing experiments: heterokaryon assays, S35 labeled protein microinjection in the nuclei of *Xenopus* oocytes and affinity chromatography using dsRBD1-2-3 of hsADAR1:

4. Discussion: 68
Studying nuclear import of hsADAR1: 69
Identification of NLS-comprising residues within dsRBD3:
Which import receptor mediates import of hsADAR1?
What is the function of the MOI? 71

5. References: 74

Acknowledgements: 98

Abstract:

The RNA editing enzyme hsADAR1 (human adenosine deaminase that acts on RNA 1) converts adenosines to inosines in double-stranded RNA. HsADAR1 is a transcription-dependent shuttling protein and expressed in two versions. The interferon inducible, longer version contains a Crm1-dependent nuclear export signal (NES) in the N-terminus and is located to both, the nucleus and the cytoplasm, where it is suggested to hyperedit viral RNAs. In contrast, the shorter, aminoterminally truncated version is constitutively expressed and predominantly nuclear. To enter the nucleus hsADAR1 contains an atypical nuclear localization signal (NLS) overlapping the third double-stranded RNA binding domain (dsRBD) in the center of the enzyme. Previous investigations addressed the question which regions of hsADAR1 contribute to regulate the nucleocytoplasmic distribution of the enzyme. We could demonstrate that dsRBD1 of hsADAR1 interferes with nuclear localization of a reporter construct containing dsRBD3 as an active NLS. Furthermore, RNA-binding is required for the import-interfering function of dsRBD1, but not for the NLS activity of dsRBD3. Finally, two competing models were developed predicting either that hsADAR1 is anchored in the cytoplasm or that dsRBD1 acts as an RNA-binding dependent NES.

This study is concentrating on the characterization of the roles of dsRBD1 and 3 in the shuttling behavior of the enzyme. On the one hand, to examine which residues within dsRBD3 are essential for NLS-karyopherin interaction, chimeric dsRBDs with and without NLS activity were constructed and tested for their ability to enter the nucleus, while sequence alignments and analysis of crystal structure data were used to reduce the number of candidate amino acids, which were finally substituted in a dsRBD without NLS activity. The fact that all chimeric and mutant dsRBDs failed to accumulate in the nucleus indicates that NLS comprising residues are spread throughout the entire dsRBD. Additionally, several karyopherins were tested in pull down and import assays whether they can interact with dsRBD3 and mediate nuclear import of hsADAR1. Recent data revealed Transportin-1 as the most probable candidate. On the other hand, to elucidate the mechanism that interferes with nuclear accumulation of hsADAR1, experiments were focused on Exportin-5, a karyopherin exporting dsRBDs and micro RNAs. Although the export factor binds ADAR1's dsRBDs in a RNA- and RanGTP dependent manner *in vitro*, cell based assays fail to confirm an involvement of Exportin-5 in the nuclear export of ADAR1.

Zusammenfassung:

Das RNA editierende Enzym hsADAR1 (homo sapiens adenosine deaminase that acts on RNA 1) desaminiert Adenosine zu Inosinen in doppelsträngiger RNA. HsADAR1 ist ein transkriptionsabhängiges Shuttling Protein, das in zwei Versionen exprimiert wird. Die längere, Interferon-induzierbare Form ist vorwiegend cytoplasmatisch und enthält ein klassisches, Leuzin-reiches Exportsignal (NES) im N-Terminus, welches in der kürzeren, konstitutiv exprimiereten Version, die hauptsächlich im Kern zu finden ist, nicht inkludiert ist. Um in den Nukleus zu gelangen, wird ein Lokalisationssignal (NLS), das in der C-terminalen der drei Doppelstrang-RNA-bindenden Domänen (dsRBD) gelegen ist, von einem Importrezeptor erkannt. Jedoch spielt auch die aminoterminale dsRBD eine wichtige Rolle in der Regulation des Kerntransports von hsADAR1. Reporterkonstrukte, die diese beiden dsRBDs enthalten, akkumulieren in einer RNA-bindungsabhängigen Weise im Cytoplasma, obwohl ein aktives NLS vorhanden ist.

In dieser Arbeit wurde die Rolle der dsRBDs in der Regulation der nukleo-cytoplasmatischen Verteilung des Enzyms analysiert. Einerseits wurde versucht, das atypische Kerntransportsignal zu charakterisieren, indem potentielle NLS-aufbauende Aminosäuren mit Hilfe von chimären dsRBD-Konstrukten, Sequenzanalysen und 3D-Strukturmodellen identifiziert wurden. Ein finales Konstrukt, das alle Kandidaten in einer ansich nicht NLS-aktiven dsRBD enthielt, wurde aber nicht in den Kern befördert. Das legt den Schluss nahe, dass alle Aminosäuren, die für NLS-Aktivität wichtig sind, über die gesamte, eine möglicherweise einzigartige Struktur aufweisende dsRBD verteilt sind. Darüberhinaus wurden verschiedene Importrezeptoren darauf getestet, ob sie an das NLS von hsADAR1 *in vitro* binden bzw. diese es in Importassays in den Kern transportieren können. Dabei stellte sich heraus, dass Transportin-1 der wahrscheinlichste Kandidat ist. Andererseits wurde untersucht, ob die N-terminale dsRBD von hsADAR1 als NES fungiert, das von Exportin-5, einem Karyopherin, das Export von RNA-bindenden Proteinen via Interaktion mit deren dsRBD vollzieht, erkannt wird. Obwohl Exp5 ADARs dsRBDs in Abhängigkeit von RNA und RanGTP binden kann, scheint der Exportfaktor unter Berücksichtigung aller Ergebnisse dennoch kein wahrscheinlicher Kandidat zu sein, hsADAR1 aus dem Kern zu transportieren.

1. Introduction:

RNA modification:

Processing of eukaryotic RNAs:

In contrast to prokaryotic mRNAs, which are already subjected to the translation machinery at the 5' end, while their 3' ends are still being transcribed, most eukaryotic RNAs obtain several modifications before they are released from the nucleus in order to fulfill their assignments.

The most striking of these alterations can be found in almost every eukaryotic pre-mRNA, which contains coding (exons) and non-coding regions (introns) after transcription. Despite the fact that every intron is flanked by short sequences showing the splice protein complex where to excise the non-coding region, not every splice site is efficiently used. Consequently, mRNAs transcribed from the same gene may vary in length and number of their fused exons. Thus, alternative splicing can theoretically provide the cell with a huge set of different proteins derived from a single gene. In fact, not every splice site is alternatively spliced but this mechanism is presumably one of the most important reasons why there are so many different proteins in different tissues in contrast to the comparably low number of genes, especially in mammals (reviewed in Orphanides and Reinberg, 2002).

For example, alternative splicing regulates the expression of sex specifc genes in the fruit fly, *Drosophila melanogaster*. The ratio between X chromosomes and autosomes defines the splicing pattern of the pre-mRNA of the *sexlethal* gene. As a consequence, the non-coding regions of the *transformer* pre-mRNA are cut out alternatively and, hence, the splice pattern of the RNA of the *double-sex* gene is determined, through which only the RNA of female flies contain the last exon (reviewed in Penalva and Sanchez, 2003).

Furthermore, mRNAs on their way to the cytoplasm are also modified at their ends providing protection against RNAses and stability but also determines their final destination, ribosome, and the rate of translation. Namely, 5' end capping occurs and at their 3' end mRNAs obtain a polyadenosine tail (reviewed in Orphanides and Reinberg, 2002).

The most recently discovered form of processing is the more subtle mechanism of RNA editing. This method of generating RNA diversity from a single gene locus includes insertion or deletion of nucleotides or their modification. In contrast to RNA splicing, removing sequences varying in their length, a single RNA editing reaction changes only one or two nucleotides (reviewed in Gott and Emeson, 2000; reviewed in Keegan et al., 2001).

RNA editing:

RNA editing can formally be defined as a minor posttranscriptional RNA processing event that alters the sequence of an RNA from that encoded in the DNA. Thus, the term includes a set of different site-specific RNA alterations.

The first discovery of RNA editing was reported, interestingly, from the mitochondria of the parasitic kinetoplastid protozoa *Trypanosoma brucei* and *Crithidia fasciculata*. It was shown that trans-acting 50-70 nucleotide guide RNAs (gRNAs) can bind to the newly transcribed mRNA of the cytochrome oxidase and template the insertion of four uridine nucleotides to correct a frameshift. Remarkably, this mechanism has been found to be absolutely essential for producing functional proteins (Benne et al, 1986). Since this first discovery, many mRNA precursors of several kinetoplast protozoa have been detected being edited by insertion or, rarely, deletion of uridine residues (Estevez and Simpson, 1999; reviewed in Gott and Emeson, 2000).

In addition, not only mRNAs are affected by RNA editing. Also transfer RNAs (tRNAs) are subjected to a large number of base modifications. Moreover, RNA editing events have also recently been noticed to take place in ribosomal RNAs (rRNAs). It could be shown that single Cs in the 16S rRNA in the mitochondria of Dictyostelium discoiduem are converted into uridines (Barth et al., 1999) and that the 7SL RNA which is part of the signal recognition particle required during protein synthesis is edited in the trypanosomatid *Leptomonas collosoma* (Ben-Shlomo et al., 1999; reviewed in Gott and Emeson, 2000).

Ribonucleotide deamination:

As mentioned above, RNA editing involves either the insertion/deletion of nucleotides or their modification. Although more recently discovered, the latter is the more widespread type of RNA editing and includes the well-characterized conversion of cytosine (C) to uracil (U) and, respectively, adenosine (A) to inosine (I) by hydrolytic deamination in mammalian nuclei as depicted in *figure 1*.

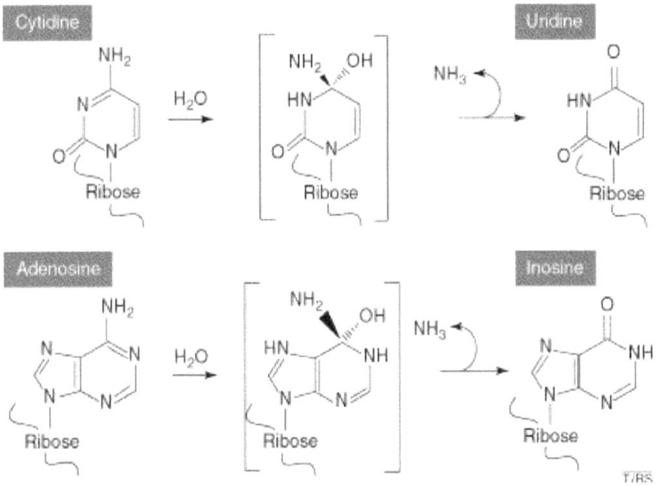

Figure 1: **Cytidine and adenosine deamination**
The reaction mechanism of cytidine and adenosine conversion includes a nucleophilic attack of an activated water molecule at the C6 of the nucleotide in an intermediate step. Subsequently, an ammonia molecule is released, resulting in a stable uridine or, respectively, inosine. (Gerber and Keller, 2001)

Cytosine deamination by APOBEC1:

The first observation of an RNA editing event in vertebrates was the specific deamination of cytosine to uridine in an mRNA encoding the mammalian intestinal apolipoprotein B (apoB) (Chen et al., 1987; Powell et al., 1987). Subsequently, this base modification has extensively been investigated and, today, it counts to one of the best-characterized examples how tissue-specific protein diversity can be generated in a simple way without changing the DNA code.

ApoB is a blood plasma lipoprotein and has been found to be essential for the metabolism of lipids. Although the protein is transcribed from one gene locus it exists in two versions in different tissues. In the liver the pre-mRNA encodes for the full-length protein, termed ApoB-100, which is part of the VLDL complex (Very-low-density lipoprotein) that is required for the transport of endogenously synthesized cholesterol and triglycerides in the blood stream. During lipid turnover, VLDL is metabolized to IDL (intermediate-density lipoprotein) and, finally, to LDL (low-density lipoprotein). Due to the fact that LDL-cholesterol tends to agglutinate in the blood, resulting in a higher risk of producing blood clots and, ultimately, of being affected by a cardiac infarct or a stroke, LDL-receptors recognize the complex specifically at the carboxyl terminus of ApoB-100 and remove it from the blood (Chan, 1992).

However, in small intestine, but not in the liver, a short version of ApoB, ApoB-48, is produced from the same gene locus but lacks the C-terminus found in ApoB-100. Particularly, this truncated protein is responsible for synthesis and secretion of chylomicrons, which are large lipoprotein complexes formed in the intestine. In contrast to other lipoproteins, chylomicrons specifically transport fatty acids from intestine membranes to the liver and to adipose tissue. ApoB-48 results from specific deamination of a cytosine at the position 6666 converted into a uridine. Consequently, a glutamine codon (CAA) is turned into a stop codon (UAA) and, thus, a shorter protein is translated.

The tissue specificity of ApoB mRNA editing is determined by expression of the cytidine deaminase APOBEC1 (ApoB mRNA-editing enzyme catalytic polypeptide 1) (Teng et al., 1993), which is the catalytic subunit of a multiprotein complex, named the editosome. Moreover, it turned out that the auxiliary protein ACF (APOBEC1 complementation factor, also known as ASF, APOBEC1 stimulating factor) is the only additional protein that is essential for the reaction (Mehta et al., 2000; Lellek et al., 2000). Although APOBEC1 can bind to the pre-mRNA of ApoB, without ACF it cannot deaminate the right cytosine. ACF contains three RNA recognition motifs (RRM) and one putative double-stranded RNA binding domain (dsRBD) and is supposed to contact the RNA ensuring specific deamination of the target cytosine. For that reason an 11-nucleotide "mooring" sequence, a spacer element and a regulatory element comprise a tripartite sequence motif, which targets the protein complex to C_{6666} (Anant and Davidson, 2001; Blanc and Davidson, 2003).

In addition, APOBEC1 has been connected with editing of the pre-mRNA of neurofibromatosis type 1 tumor suppressor (NF1). In fact, in vitro studies confirmed that APOBEC1 could edit NF1 pre-mRNA similarly to ApoB's pre-mRNA (Chester et al., 2000). Furthermore, since the discovery of APOBEC1, other members of the cytidine deaminases that act on RNA (CDAR) family have been identified in mammals. Among these, activation-induced cytidine deaminase (AID) has extensively been investigated because it is required for somatic hypermutation during immunoglobulin class switch in B lymphocytes (Muramatsu et al., 2000; Blanc and Davidson, 2003; reviewed in Keegan et al., 2001).

Adenosine deamination by ADATs:

Similar to cytidine deamination, adenosines can be converted into inosines (*Fig. 1*). Despite the fact that inosines do not naturally occur in RNAs without enzymatic interference, these nucleotides have already been found in tRNAs in 1965 (Holley et al., 1965). Since then, inosines have also been detected in mRNAs where they can make up to 9% of an RNA molecule, for example in human brains.

Due to the fact that inosines are present in the first anticodon position (wobble position 34, A34) in eight different tRNAs of higher eukaryotes, in seven different in yeast and in tRNA^{Arg2} of prokaryotes and chloroplasts in plants, it is assumed that inosines play an essential role in protein synthesis. Regarding to the triplet code, these inosines allow alternative base pairs with U, C or A in the third codon position.

Additionally, N^1-methylinosine (m^{1}I$_{37}$) has been detected at position 37 of tRNA^{ala} (A37) adjacent to the anticodon. Although deamination of A37 of this single tRNA turned out to be not absolutely essential in yeast, m^{1}I$_{37}$ probably prevents translational frameshifts and therefore ensures correct protein synthesis (Gerber and Keller, 1999).

Data base search in yeast disclosed an open reading frame that showed homology to already discovered adenosine deaminases in the C-terminal end. Indeed, it turned out that the enzyme, termed Tad1p (tRNA specific adenosine deaminase) or scADAT 1 (*Saccharomyces cerevisiae* adenosine deaminase that acts on tRNA), targets exclusively A37 of tRNA^{ala} (Gerber et al., 1998). Subsequently, ADAT1 was cloned from humans (Maas et al., 1999), mouse (Maas et al., 2000) and from *Drosophila melanogaster* (Keegan et al., 2000). In contrast to other adenosine deaminases, ADAT1 does not contain an RNA binding motif indicating that the deaminase domain directly recognizes its substrate.

A few years ago two more deaminases have been described in yeast (Tad2p/scADAT2 and Tad3p/scADAT3), which are capable of triggering the conversion of A34 into an inosine. Interestingly, ADAT2 seems to be the catalytic subunit of a heterodimer formed by the two proteins. Remarkably, in contrast to ADAT1, knock out experiments in yeast revealed that ADAT2$^-$/ADAT3$^-$ strains are lethal underlining the importance of this editing mechanism (Gerber et al., 1999; reviewed in Gerber and Keller, 2001; reviewed in Keegan et al., 2001). In addition, an ADAT2 homologue, tadA, which edits tRNA^{Arg2} at position 34, could have been detected in the prokaryote *Escherichia coli* (Wolf et al., 2002).

ADARs (adenosine deaminases that act on RNA):

Discovery of ADARs:

In 1987, two laboratories observed almost simultaneously that double-stranded RNAs when injected into *Xenopus laevis* oocytes experience changes in their structure. Interestingly, they could detect mobility shifts on native RNA gels and an altered susceptibility to single-strand specific RNAses (Bass and Weintraub, 1987; Rebagliati and Melton, 1987). Subsequently, the same observations could have been done in mammalian cells (Wagner and Nishikura, 1988) and were postulated as an unwinding activity.

However, it was later shown that "unwinding" resulted from the covalent modification of adenosines to inosines. Hence, A-U base pairs are converted into I-U mismatches which destabilizes the RNA duplex (Bass and Weintraub, 1988; Wagner et al., 1989; Polson et al., 1991). The responsible enzyme was first found in frogs (Hough and Bass, 1994) and shortly after in mammals (Kim et al., 1994; O'Connell and Keller, 1994).

Due to the fact that the same RNA altering activity was identified in different laboratories, two names for the same enzyme, dsRAD (<u>ds</u>RNA <u>a</u>denosine <u>d</u>eaminase) and DRADA (<u>ds</u>RNA <u>a</u>denosine <u>d</u>eaminase), were published. In 1996, Melcher and colleagues identified a second enzyme, RED 1 (ds<u>R</u>NA specific <u>ed</u>itase 1) with such a RNA editing ability (Melcher et al., 1996). However, to avoid further complications a committee standardized the names to ADAR 1 and ADAR 2 (Bass et al., 1997).

Substrate recognition and substrates of ADARs:

Together with ADATs, ADARs comprise a protein family that edits adenosines in structured or double-stranded RNAs. If the edited RNA is an mRNA the observed base modification can result in a codon exchange as inosines are interpreted as guanosines during translation or it can affect pre-mRNA splicing by introducing or destroying splice sites. The fact that perfect double-stranded RNA (dsRNA) becomes predominantly hyperedited where up to 50% of the adenosines on each strand are affected in an almost random pattern (Nishikura et al., 1991; Polson and Bass, 1994) together with the finding that ADARs are interferon inducible suggests the enzymes participate in the cellular antiviral defense system. Thus, it is assumed that ADARs are fighting viruses that use dsRNA in their replication cycle (reviewed in Bass, 2002, Keegan et al., 2001). On the other hand, specific editing affecting only some nucleotides mostly occurs in endogenous RNAs. In principal, the more internal loops and bulges a double-stranded, structured RNA contains, the more likely specific editing takes place. In fact, the optimal binding selectivity is obtained when a short RNA helix flanked by unpaired regions is involved. Included mismatches destabilize the double-strand what stops the reaction automatically when the two strands can no longer pair due to further disruption caused by adenosine deamination (Polson and Bass, 1994; Lehmann and Bass, 2000; reviewed in Bass, 2002).

Currently, it is poorly understood how the enzyme discriminates between adenosines that have to be converted into inosines and those that are not altered. It appears likely, however, that ADARs recognize their substrates specifically on their structure but not on their sequence. As suggested for ADATs, substrate recognition by ADARs seems to be achieved by an interplay between double-stranded RNA binding domains (dsRBDs) and the deaminase domain (Källmann et al., 2003; Wong et al., 2001; reviewed in Bass, 2002; reviewed in Wang and Carmichael, 2004).

Editing of viral dsRNA:

As mentioned above, dsRNA using viruses are often subjected to hyperediting and since the discovery of ADARs, a couple of viruses have been detected being edited. Among these are the Hepatitis delta virus (Polson et al., 1996), the measles virus (Cattaneo et al., 1989), the polyoma virus (Kumar and Carmichael, 1997), but also the avian retrovirus ALV (Hajjar and Linial, 1995). In fact, although not completely understood, ADAR1 might play a role in limiting replication of the hepatitis C virus in humans (Taylor et al., 2005).

Generally, it is assumed that ADAR1 edits viral dsRNA in the nucleus and in the cytoplasm and, thus, it is becomes partly single stranded, which makes it more prone to digestion by ribonucleases. Indeed, the existence of an RNAse explicitly cleaving these RNAs has been proposed for years. Recently, a cytoplasmic endonuclease activity that specifically targets promiscuously edited dsRNA, whereas non-edited RNA is not affected, was described (Scadden and Smith, 1997; Scadden and Smith, 2001; Wong et al., 2003). On the other hand, abundant inosine containing RNAs are bound and retained by the Vigilin protein family in the nucleus. But, RNAs with only one or a few inosines can escape nuclear retention and are transported to the cytoplasm (Kumar and Carmichael, 1997; Zhang and Carmichael, 2001; Wang et al., 2005).

To date, there is only one example of a virus that takes advantage of adenosine deamination in its life cycle. ADARs deaminate a single A in the only open reading frame (ORF) of the hepatitis delta virus to convert a stop codon into a tryptophane. Hence, two different proteins are generated which are both essential for the viral life cycle. The short product is required for viral replication, whereas the longer form is needed for assembly of new viral particles (Polson et al., 1996; Kuo et al., 1989; Chang et al., 1991). However, when editing is inhibited, viral particles cannot assemble and, thus, the virus is no longer infectious (reviewed in Taylor, 1999).

Endogenous substrates: serotonin receptor 5-HT$_{2C}$R

While it is estimated that up to 10% of mRNAs might be edited in some tissues (Paul and Bass, 1998), only a few substrates of human ADAR1 (hsADAR1) are currently known. Consistent with the rate of ADAR's expression and the level of inosine containing RNAs, is the fact that the first identified specifically edited substrates were found in neuronal tissues. The best-characterized examples are the pre-mRNAs encoding subunits of the glutamate receptor, the serotonin 2C receptor in mammals or the potassium channel in squids (Burns et al., 1997; Lehmann and Bass, 2000; Melcher et al., 1996; Patton et al., 1997).

Serotonin is a neurotransmitter important for neuronal functions such as sleep regulation, appetite or pain. The molecule interacts with a large family of receptors and triggers signaling that is crucial for proper neurotransmission. Indeed, binding of serotonin to its transmembrane receptor is involved in several physiological and behavioral processes such as production of cerebrospinal fluid and regulation of feeding behavior (Maas and Rich, 2000). As can be imagined, these receptors are implicated in several psychiatric disorders such as schizophrenia, but are also targeted by antipsychotic drugs, for example lysergic acid diethylamide (LSD) (Niswender et al., 2001).

Interestingly, in rat at least seven and in humans twelve different isoforms of the 2C subtype of serotonin receptor (5-HT$_{2C}$R) have been found resulting from alternative editing at five different sites by ADARs (Burns et al., 1997; Niswender et al., 1998). Strikingly, the resulting amino acid substitutions affect ligand affinity, by which the unedited version exhibits the highest affinity to the neurotransmitter. Furthermore, recent investigations revealed a connection between serotonin levels and the extent of editing of the 5-HT$_{2C}$R mRNA. Therefore, it is suggested that editing participates in a self-regulatory mechanism keeping the receptors in an optimal range of activation in a serotonin dose dependent manner (Gurevich et al., 2002).

As in several examples of ADAR substrates, the editing sites occur within dsRNA that involves base pairing between an intron and an exon. Here, five adenosines can be deaminated that are almost adjacent to each other within the pre-mRNA encoding the second intracellular loop of 5-HT$_{2C}$R (Niswender et al., 1998).

Another well-characterized example is editing of the glutamate receptor subunit B (GluR-B) (Sommer et al., 1991). Glutamate receptors assemble from related subunits and mediate fast neurotransmission in the brain by increasing the Ca^{2+} concentration within the cells. ADARs modify the RNAs of GluR-Bs at three sites, two of which are located within exons and therefore cause codon exchanges. At the Q/R site a glutamine (Q) is converted into an arginine (R) at almost 100% by ADAR2. Interestingly, this editing event turned out to be absolutely essential for the function of the glutamate receptor (Brusa et al., 1995; Dabiri et al., 1996).

The second site within GluR-B altering a triplet, affects an arginine that is turned into a glycine (R/G site). Deamination at this site can be done by both, ADAR1 and ADAR2, and its efficiency rises during brain development. In this case editing modifies the channel kinetics of the receptor (Lomeli et al., 1994).

New substrates: Alu elements, UTRs, self-editing of ADAR, miRNAs:

Due to new approaches in comparative genomics and expressed sequence analysis, a number of new endogenous substrates have recently been identified. Interestingly, the majority of editing events displays a non-selective pattern and can be found in untranslated regions (UTRs) and introns of mRNAs where large duplexes are formed. Especially Alu elements that abundantly occur in such non-coding regions are often hit by deamination.

Alu elements are approximately 300 base pairs in length and contain inverted repeats that, unless mutated, can fold into perfect double-strands. Therefore, it is assumed that the majority of all mRNAs represent potential substrates for ADARs (Morse et al., 2002; Athanasiadis et al., 2004; Levanon et al., 2004).

In addition, Levanon and colleagues also detected new selectively edited substrates. Fascinatingly, the finding that none of these substrates encodes a receptor protein together with the fact that in this approach not all of the already known substrates were found, gives rise to the assumption that there are still a lot undetected substrates (Levanon et al., 2005).

An example for creating a new splice site by editing is mammalian ADAR2, acting on its own mRNA. Thereby, an AA dinucleotide is converted into AI, which is recognized as AG typically found at 3' splice sites. Generation and usage of this newly created splice site results in a frameshift and the production of a truncated protein. Presumably, this process is part of an auto-regulatory mechanism, which tightly sets the levels of ADAR2 (Rueter et al., 1999).

Recently, a new class of non-coding RNAs that are involved in RNA expression regulation was shown being affected by adenosine deamination. The precursor of micro RNA 22 (miRNA 22) forms an 85 nucleotide stem-loop that can be bound and edited at up to six sites by ADAR1 and 2 in humans. Currently, it is poorly understood for what purpose miRNA 22 is edited but it seems to fine tune the function of the miRNA by modulating the interaction with its target sequence (Luciano et al., 2004).

On the other hand, it appears likely that ADARs' activity antagonizes the RNAi mechanism. The formation of partly single stranded structures makes the RNA less attractive for the RNAi machinery. Hence, ADARs may decide whether a dsRNA enters the RNAi pathway or not (Scadden and Smith, 2001; Knight and Bass, 2002; Tonkin and Bass, 2003).

Biological importance of ADARs:

ADARs can be found in all metazoans and related enzymes have been cloned from several organisms, including *C. elegans, D. melanogaster, X. laevis* and also mammalian species including humans (Bass, 1997). It appears, that ADARs have gained more and more functions during evolution and have meanwhile proven to be essential enzymes.

In contrast to mammals, *C. elegans and D. melanogaster* strains that lack ADAR activity are viable. However, they exhibit mostly developmental and behavioral defects regarding to the high expression level of ADARs in neuronal tissues. For example, *C. elegans* has two ADAR genes, adr-1 and adr-2, with the first being expressed in most cells of the nervous system and developing vulva. Knock out experiments showed that worms lacking ADAR activity have severe defects in chemotaxis. Additionally, both adr-1 and adr-2 are important for normal behavior (Tonkin et al., 2002).

However, higher organisms completely missing the activity of ADARs are not viable. In fact, embryonic lethality on day 12.5 and defects in the hematopoietic system are observed in mice lacking one single allele of ADAR1 (Wang et al., 2000). In contrast to ADAR1, heterozygote ADAR2 $^{+/-}$ mice are viable. However, homozygote ADAR2 $^{-/-}$ mice die short after birth having extensive brain seizures developed. Remarkably, by substituting the glutamine by an arginine at the Q/R site of GluR-B in ADAR2 knock-out mice, the phenotype can be rescued (Higuchi et al., 2000).

Furthermore, ADAR1 heterozygosity in mice was reported to cause defects in the hematopoietic system and leads to embryonic lethality before day 14 (Wang et al., 2000). In ADAR1 $^{-/-}$ mice embryos, however, severe liver defects can be detected, cells in diverse tissues undergo stress-induced apoptosis and the animals die between day 11.5 and 12.5 (Hartner et al., 2003; Wang et al., 2004).

Finally, a number of human diseases have been connected with ADAR's editing activity. Besides several psychiatric disorders such as depression and schizophrenia depending on different serotonin receptor isoforms, as mentioned above, also physical illnesses seem to be caused by mutated or misregulated adenosine deaminases. For example, not entirely edited Q/R-site in the pre-mRNA of GluR-B due to a reduced enzymatic activity of ADAR2 correlates with the occurrence of patients developing malignant gliomas, indicating a role for RNA editing in tumor progression (Maas et al., 2001). Furthermore, the skin disorder Dyschromatosis symmetrica hereditaria (DSH) was recently associated with two mutations in ADAR1 (Liu et al., 2004).

In summary, the diversity of phenotypes A-to-I-conversions can generate together with the finding that at least ADAR1 participates in the antiviral defense mechanism and might fine-tune the important RNAi machinery in mammals, underscores the assumption that ADARs are key enzymes that have to be tightly regulated.

Members of the human ADAR family:

The hsADAR1 gene locus contains a constitutive and an interferon-inducible promotor and, thus, two versions of the protein differing in their amino-terminal ends are produced. The approximately 110 kDa protein transcribed from the constitutive promotor, also termed hsADAR1c, starts at methionine 296 and lacks the first exon (George and Samuel, 1999; Patterson and Samuel, 1995). In contrast, the interferon-inducible variant, termed hsADAR1i, starts at methionine 1 and can be mostly found in the cytoplasm where the 150 kDa protein has recently been implicated in editing of viral RNAs (Wong et al., 2003). Additionally, a proteolytical cleavage site has been proposed to occur in ADAR1 (Patterson and Samuel, 1995; Eckmann and Jantsch, 1999) and alternatively spliced versions of hsADAR1 have been detected (Liu et al., 1997; George et al., 2005), contributing to diversify the different isoforms of ADAR1 in a cell in order to fulfill their various functions.

As depicted in *figure 2*, ADAR2 shows an enormous structural similarity to ADAR1. The approximately 80 kDa protein has 31% sequence identity to ADAR1 with the C-terminal catalytic domain being most conserved (Melcher et al., 1996). Importantly, ADAR2 contains only two dsRBDs and no Z-DNA binding domains (ZBDs) (see below). ADAR1 is expressed in most tissues, which is in agreement with the observation that mRNAs containing inosines can be found in the majority of cell types. ADAR2, in contrast, is predominantly expressed in the brain, which correlates with the highest abundance of inosines containing RNAs in the brain (Paul and Bass, 1998).

The most recently discovered member of the human ADAR family, ADAR3, is about 81 kDa in size, has been found being exclusively expressed in the brain, particularly in the amygdala and the thalamus (Mittaz et al., 1997), and shows 50% sequence identity to ADAR2. ADAR3 contains two dsRBDs and a conserved deaminase domain. Additionally, it further contains a 54 amino acid arginine-lysine-rich sequence (R-domain) in the N-terminus, which has been identified as a single-stranded RNA binding domain (Chen et al., 2000). However, no substrate of ADAR3 has been identified so far suggesting that the enzyme lacks catalytic activity at all. In fact, ADAR3 is unable to edit any of the known editing sites of ADAR1 and 2 or adenosines in an artificial substrate. Further experiments indicate that ADAR3 binds potential substrates of ADAR1 and 2 and, thus, inhibits editing of the two deaminases (Chen et al., 2000). An alternative model suggests that heterodimerization of ADAR3 with either of the two other family members might also decrease editing activity as dimerization might be necessary for catalytic activity.

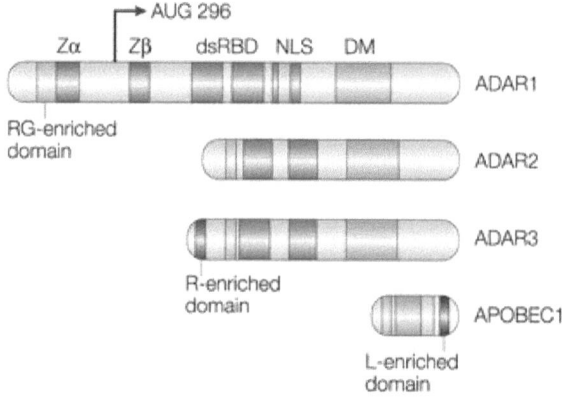

Figure 2: **Schematic diagram of human ADAR and APOBEC 1:**

Common in all adenosine and cytidine deaminases is the conserved deaminase domain (DM) in the C-terminus. In contrast to APOBEC1, ADARs contain double-stranded RNA-binding domains in the center of the proteins that are important for substrate recognition and RNA binding. Additionally, only ADAR1 harbors two Z-DNA binding domains (Zα and β) and an arginine-glycine rich domain (RG) in the N-terminus. In ADAR2 an arginine rich (R) and in APOBEC1 a leucine rich domain (L) can be found. To enter the nucleus, ADAR1 contains a nuclear localization signal (NLS) overlapping the third dsRBD, whereas in ADAR2, 3 and APOBEC1 it is located in the N-terminal end of the protein. A methionine codon (AUG 296) from which translation of ADAR1 is alternatively initiated is illustrated by an arrow (Keegan et al., 2001).

Functional domains of ADAR:

As stated above, all ADAR and ADAT proteins contain a conserved deaminase domain in their C-terminus but differ in their central and aminoterminal regions. This domain bears high homology to the catalytic center of some DNA methyl-transferases and CDARs (cytidine deaminases that act on RNA) (Hough and Bass, 1997; Kim et al., 1994).

ADAR's deaminase domain is composed of characteristic sequence motifs involved in the catalysis of hydrolytic deamination. One motif includes a glutamate, which is assumed to be required for mediating proton transfers. Additionally, conserved histidine and cysteine residues complex a zinc ion and, thus, comprising the active center. As shown for other metalloenzymes, zinc activates a water molecule and therefore its oxygen can nucleophilicly attack the adenosine at its C6 (Lai et al., 1995). Similar to the catalytic mechanism of methyl-transferases, the affected adenosine is flipped out of the RNA helix due to spatial reasons during the editing reaction (Stephens et al., 2000). Interestingly, the dsRBDs of ADARs seem to support this process by enhancing the flexibility of the RNA duplex (Yi-Brunozzi et al., 2001).

In addition to the deaminase domain, only the interferon inducible version of human ADAR1 harbors two Z-DNA binding domains (ZBD α and β) in its N-terminus. Investigated by NMR and crystal structure analyses, the approximately 65 amino acids of each domain build up a helix-loop-helix structure, similar to DNA-binding proteins, and can bind left-handed, supercoiled Z-form DNA *in vitro* (Schade et al., 1999; Schwartz et al., 1999; Herbert et al., 1997). However, Z-DNA has not been found *in vivo* so far, but for energetically reasons it might occur directly behind active RNA polymerases. Therefore, it has been speculated that the ZBDs directs ADAR1 to transcriptional active sites (Herbert al., 1997). On the contrary, it was shown for *Xenopus laevis* ADAR1 that the ZBDs are dispensable for chromosomal localization (Eckmann and Jantsch, 1999). Finally, different models suggest that ZBDs are important for binding of viral dsRNA in the cytoplasm (Brown et al., 2000; Herbert et al., 1997; Kim et al., 2003) and additionally they seem necessary to recruit small substrate RNAs (Herbert and Rich, 2001).

In the central region ADAR1 contains three double-stranded RNA-binding domains (dsRBDs), which are seemingly involved in substrate recognition (see below). ADAR 2 and ADAR 3 contain only two dsRBDs.

Double-stranded RNA binding domains (dsRBDs):

ADARs belong to the dsRNA binding proteins (DRBPs) that share a common, evolutionary conserved motif specifically facilitating interactions with dsRNA. Since the first was discovered in

1992, so far, more than 20 DRPBs have been identified from bacteria to mammals with partly critical roles with the cell (St Johnson et al., 1992). Among these is Dicer, for instance, which contains only one dsRBD and is essential in the RNAi mediated gene silencing mechanism. Another example is the interferon-inducible dsRNA-dependent protein kinase PKR that is implicated in dsRNA signaling and host defense against viral infection. Evidently, dsRBDs are successful structures that enable the proteins to bind diverse RNA, but on the other hand, in addition, they can also have surprising functions.

DsRBDs are typically 65-70 nucleotides in length and form three β-sheets enclosed by two α-helices. In fact, the helices lie on the face of three antiparallel β-sheets (Bycroft et al., 1995; Kharrat et al., 1995). NMR and crystallographic studies revealed that dsRBDs interact with RNA at its sugar-phosphate backbone without any direct base contact covering 11-16 nucleotides. Thereby, the A-form RNA helix is slightly bent and forms deep and narrow major grooves and shallow minor grooves in contrast to B-form DNA (Ryter and Schultz, 1998). Consequently, dsRBDs seem to bind RNA without any sequence specificity (reviewed in Barber and Saunders, 2003; reviewed in Tian et al., 2004). Thus, the key questions have arisen: how can dsRBDs discriminate between different substrates and how can ADARs edit them selectively or non-selectively?

The answers to these questions are still not completely solved but it appears likely that the fact that several DRBPs contain more than one dsRBD helps to clarify this problem. Actually, most DRBPs contain multiple, tandemly arranged dsRBDs and it is believed that they contribute to extend the range of potential substrates of a protein. That notion was supported by the recent finding that each dsRBD of ADAR2 can bind distinct RNA-structures at the Q/R-site of GluR-B mRNA, meaning that they contribute for sequence-independent but structure-selective binding (Carlson et al., 2003; Stephens et al., 2004; Stefl et al., 2005). On the other hand, the deaminase domain was shown not only participating in the RNA discrimination process, it is the main determinant of specificity. This was confirmed by chimeric ADAR proteins, in which the deaminase domains between ADAR1 and 2 were exchanged (Wong et al., 2001).

However, as different dsRBDs have different affinities for dsRNAs (Krovat and Jantsch, 1996) it seems that the number and kind of dsRBDs within a protein is important for its function. A well-characterized example is Staufen of *Drosophila melanogaster*, which could have been implicated in several mechanisms of fly embryogenesis. The protein contains five dsRBDs and each of which has different functions in the localization of mRNAs important for the development of the flies (Micklem et al., 2000).

Another interesting point is that some dsRBDs, not only in Staufen, are even incapable of binding RNA in vitro (Krovat and Jantsch, 1996). This finding can be explained by the discovery of dsRBDs fulfilling additional functions except RNA binding. Seemingly, the structure allows dsRBDs not only to bind RNA but also to interact with other cellular components, even with other dsRBDs. For instance, xlRBPA (*Xenopus laevis* RNA-binding protein A) can bind the majority of cellular RNAs, including ribosomal RNAs and hnRNAs, but *in vitro* binding experiments revealed that only the second dsRBD of this protein is able to interact with an RNA as an isolated domain. Additionally, it was shown that the highly conserved dsRBD3 of XlrbpA and its human homologue, PACT, homodimerize in an RNA independent manner (Eckmann and Jantsch, 1997; Hitti et al., 2004).

In addition, among other dsRBD containing proteins, such as PKR, Staufen or RNAse III, also ADARs have been disclosed to perform multimerization (Cosentino et al., 1995; Wu and Kaufman, 1997; Patel and Sen, 1998; Romano et al., 1998; Lamontagne et al., 2000). The N-terminus and the first dsRBD of *Drosophila* ADAR1 were demonstrated to be essential for RNA editing of this protein because the enzyme is only active as a dimer and these regions are required for dimerization (Gallo et al., 2003). Even human ADAR1 and 2 are suggested homodimerizing to comprise to an enzymatically active complex. Thereby, the monomeric ADAR3 is thought to act as a modulator of dimerization of its relatives in order to regulate the activity of ADAR1 and 2 in the brain (Cho et al., 2003). However, this notion still remains an open question in the ADAR community.

Another surprising function of some dsRBDs is their role in interacting with the nucleocytoplasmic transport machinery. By investigating which region of hsADAR1 mediates nuclear import, it was observed that the third dsRBD is sufficient and essential for movement of the protein into the nucleus. Although it has been previously known that this dsRBD is most important for RNA binding (Lai et al., 1995; Liu and Samuel, 1996), the nuclear localization signal (NLS) activity was found to be independent of RNA-binding (Eckmann et al., 2001). Furthermore, dsRBD2 of the interleukin enhancing factor 3 (ILF3) serves as a nuclear export signal (NES) that is recognized by the export factor exportin5 (Exp5) in the presence of bound RNA (Brownawell and Macara, 2002; Gwizdek et al., 2004) (see below).

In contrast to nuclear export, an RNA dependent cytoplasmic anchoring function has been detected for the dsRBDs of the Xenopus CCAAT box transcription factor (CBTF) (Brzostowski et al., 2000). On the other hand, different dsRBDs localize *Xenopus* ADAR1 to different transcriptional active loops on lampbrush chromosomes in oocytes (Doyle and Jantsch, 2003). In the absence of transcription, however, the enzyme strongly accumulates at one single loop, termed the "special loop" (Sallacz and Jantsch, 2005).

Nuclear transport:

Eukaryotic cells are characterized by the existence of a double-membraned nuclear envelope, which separates the nucleus from the cytoplasm. Transport between these compartments occurs through nuclear pore complexes (NPCs), which are multiprotein complexes, comprising aqueous channels. In contrast to small molecules and ions that can pass the NPC by free diffusion, particles that are larger than 30-40 kDa have to use a signal- and energy-dependent pathway in order to translocate to the other side of the NPC. For that reason, nuclear localization signals (NLSs) or nuclear export signals (NESs) within the cargo proteins are recognized by soluble transport factors, also known as karyopherins, which interact with the NPCs. Other large molecules, such as RNAs, have to associate with proteins containing a transport signal to change the side of the NPC. To impose the directionality of nucleocytoplasmic traffic, the small GTPase Ran can uptake and hydrolyze a GTP and thereby fuels this process (reviewed in Nakielny and Dreyfuss, 1999).

The Nuclear Pore Complex (NPC):

In the middle of the last century, for the first time pores in the nuclear envelope of eukaryotic cells were discovered by electron microscopy studies (Callan and Tomlin, 1950). Although the characteristic eightfold structure of the pores were first described shortly after, leading to the name NPC (Watson, 1959), it took many years to reveal the pores being multiprotein complexes (Aaronson and Blobel, 1974).

Today, still not all components are identified, but it is clear that the NPC is a universal feature of the nuclear envelope in all eukaryotes and the physical structure has been resolved. Depending on the size and type of a cell, between 3000 and 5000 NPCs can be found embedded in the nuclear membrane and each transfers up to 1000 molecules per second (Ribbeck and Görlich, 2001). The NPC is composed of more than 50 proteins in yeast and more than 100 in humans, resulting in a total mass of 90-120 MDa (Cronshaw et al., 2002). The protein complexes form three main structural parts that include the central framework, the cytoplasmic ring with filaments and the nuclear basket (Stoffler et al., 2003). The eightfold structure allows the pore a particular flexibility, which is necessary considering the fact that molecules of all different sizes have to pass in a controlled way. Gold-particles microinjection experiments with *Xenopus* oocytes disclosed that particles up to 26nm in diameter can be transported by the NPC (Dworetzky et al., 1988). However, a more recent study concluded that the NPC is able to transport molecules with diameters of up to 39nm (Pante and Kann, 2002; reviewed in Fahrenkrog and Aebi, 2003).

The approximately 30 different proteins in yeast and humans comprising an NPC have been termed nucleoporins (Nups) (Rout et al., 2000; Cronshaw et al., 2002). Most of them contain coiled-coil or leucine-zipper motifs enabling interaction among each other, and roughly one third of Nups contain phenylalanine-glycine repeats (FG-repeats) which are bound by transport factors during nucleocytoplasmic transfer (Rexach and Blobel, 1995; Shah et al., 1998). For example, Nup50 can interact with import factors and escorts cargos from the nucleus to the cytoplasm and vice versa. Thereby, the interaction involves primarily the phenylalanine ring of the FG-repeats and hydrophobic residues on the surface of the receptors (Lindsay et al., 2002).

Summarizing, the NPC is a very flexible and dynamic protein complex that regulates nuclear transport selectively and efficiently maintaining the separation of the two compartments but also providing rapid translocation of molecules.

The RanGTPase cycle:

As shown for many transport mechanisms between two cellular compartments, a nucleotidase, in case of nucleocytoplasmic transfer, the small GTPase Ran, determines the direction of the transport and supplies the process with energy. Due to its abilities to hydrolyze GTP to GDP + Pi and to interact with transport factors, it affects the affinity of karyopherins to their freight and, thus, the directionality of the movements. Thereby, Ran shuttles between to nucleus and the cytoplasm allowing both, import and export of cargo proteins (reviewed in Nakielny and Dreyfuss, 1999).

In fact, the directionality of transport is exclusively determined by a RanGTP gradient, and not by the asymmetry of the NPC as previously assumed (Görlich et al., 2003), maintained by several proteins stationary in either of the two compartments. Import factors recruit their NLS containing cargos at low RanGTP concentrations in the cytoplasm and cross the nuclear envelope without any further requirements. But to release its freight, RanGTP interacts with the import factor and remains bound. This newly formed complex returns afterwards back to the cytoplasm, where Ran-bound GTP is hydrolyzed by Ran supported by the RanGTPase activating protein (RanGAP) (Bischoff et al., 1994) and by the Ran binding proteins 1 and 2 (RanBP1 and 2) (Bischoff et al., 1995; Yokoyama et al., 1995). Subsequently, the heterodimer dissociates and the import factor can bind and import another cargo molecule, while the Nuclear Transport factor 2 (NTF2) recycles RanGDP back to the nucleus (Ribbeck et al., 1998; Smith et al., 1998). Maintaining the gradient, the Ran guanine exchange factor (RanGEF), also termed Ran-specific nucleotide exchange factor (RCC1), is associated with chromatin in the nucleus and replaces GDP by GTP bound by Ran (Ohtsubo et al., 1989; Bischoff and Ponstingl, 1991; Smith et al., 2002).

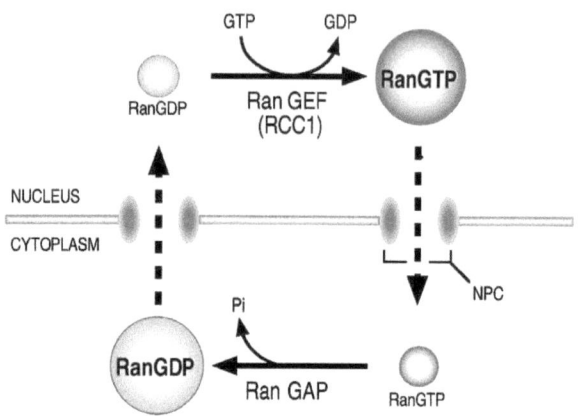

Figure 3: **The RanGTPase cycle:**
The RanGTP concentration in the nucleus is maintained by exchanging Ran-bound GDP to GTP by RanGEF (RCC1). On the other side of the nuclear envelope after export via a trimeric complex, GTP hydrolyzation is facilitated by Ran and supported by RanGAP, RanBP 1 and 2 (Nakielny and Dreyfuss, 1999).

Similarly, nuclear export is controlled in the opposite manner. A cargo-export factor complex can only escape from the nucleus when bound to RanGTP. In the cytoplasm Ran hydrolizes GTP to GDP, supported by RanGAP, RanBP1 and 2, and the trimeric complex dissociates. Consequently, the export factor and RanGDP move separately back to the nucleus (Kutay et al., 1997; Görlich, 1998; Kehlenbach et al., 1999).

In addition, Ran also plays an essential role in mitosis. The nucleotidase controls the assembly of mitotic spindles by relieving binding of certain microtubule-associated proteins to the import factors importin α and β. Thus, spindle assembly is promoted in close proximity to chromatin, where RanGTP is maximal concentrated, as a result of RCC1's nucleotide exchange activity (Kalab et al., 1999; Gruss et al., 2001; Nachury et al., 2001). Furthermore, during nuclear envelope assembly the RanGTP gradient provides positional information, ensuring that the newly built nucleus encloses the whole chromatin (Hetzer et al., 2000; Zhang and Clarke, 2000).

Transport factors, cargos and their localization signals:

Transport of proteins and ribonucleotide particles (RNPs) across the nuclear membrane is specifically mediated by a group of receptor proteins that bind their cargo via a particular sequence, transport it through the nuclear pore complex and release it at the other side. These sequences are termed nuclear localization signals (NLS) in case of import and nuclear export signals (NES) for export. Dependent on the direction of transport the soluble factors are typically referred to as importins or exportins but are also commonly called karyopherins.

The vast majority of these receptors belong to a protein family, named the importin β family referring to one of the first and best characterized transport factors. Currently, there are at least 24 members known that share homology with this karyopherin. Most of the members vary between 90 and 130 kDa and have an acidic isoelectric point. Only the N-termini of the receptors are conserved which are responsible for interactions of the karyopherin with the NPC and RanGTP, in contrast to the cargo recognition sites in the C-terminal halves which, in most cases, do not even exhibit any sequence similarities (Kutay et al., 1997; Chi and Adam, 1997; reviewed in Ström and Weis, 2001).

Consistently, the diversity of transport signals includes from short, basic and unstructured amino acid stretches to functional domains that contain multiple short sequences recognized by a transport receptor (reviewed in Mattaj and Englmeier, 1998; reviewed in Nakielny and Dreyfuss, 1999). The variety of transport pathways will be illustrated by three examples.

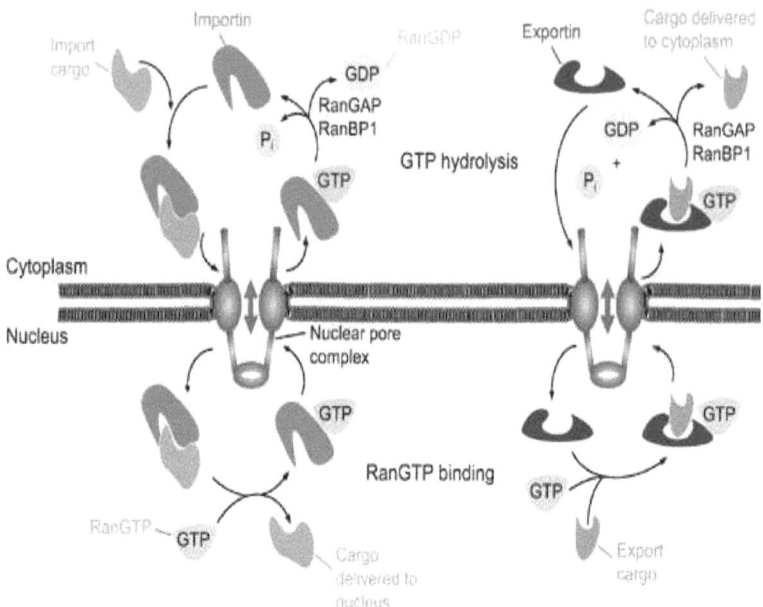

Figure 4: **Import and export cycles:**
Import receptors take up their cargo in the cytoplasm and translocate through the NPC. In the nucleus binding of RanGTP triggers release of the NLS containing protein and the new assembled complex turns back to the cytoplasm where it does not dissociate unless RanGAP and RanBP1 stimulate GTP hydrolyzation. In contrast, exportins recognize their cargoes in the nucleus and form a trimeric complex with RanGTP. After transfer to the cytoplasm, GTP is hydrolyzed what reduces the affinity of Ran to the exportin. Consequently, the complex disassembles and the export receptor and RanGDP move separately back to the nucleus (Ström and Weis, 2001).

The Importin α/β import pathway:

NLS recognition followed by transport through the NPC by importin α and importin β (Impα andβ) has been extensively studied and, thus, represents a key example of nuclear import. In contrast to many other receptors, Imp β does not bind its freight directly. First cloned from *Xenopus* (Görlich et al., 1994), Impα, however, acts as an adaptor, binding the substrate protein by recognition of the NLS with a domain in its C-terminus (Conti et al., 1998; Herold et al., 1998).

Interestingly, Impα harbors an auto-inhibitory domain that blocks the NLS-binding site unless Impβ has not bound to the importinβ-binding domain (IBB) of the adaptor protein (Kobe, 1999; Fanara et al., 2000). Once the trimeric complex has formed, a region in the N-terminus of Imp β (see Fig. 5) immediately interacts with the FG repeats of several nucleoporins and the complex can pass the NPC (Chi and Adam, 1997; Shah et al., 1998; Bayliss et al., 2000). In the nucleus, RanGTP removes Impβ from the complex (Görlich et al., 1996) and the Impα-cargo heterodimer dissociates, assisted by binding of the karyopherin CAS to the adaptor protein (Kutay et al., 1997).

Currently, there are six subtypes of Impα in mammals known that can interact with different NLSs. Considering the fact that the importinα/β pathway represents one of the major player of the import machinery, it makes sense to produce several adaptors for one actual receptor.

The most investigated import signals using this pathway are classical NLS of the large T antigen of the simian virus 40 (SV40) and the bipartite NLS, first identified in nucleoplasmin (Kalderon et al., 1984; Robins et al., 1991). In the first case the NLS consists of a single cluster of basic amino acids (lysines and arginines), often preceded by an acidic or a proline residue. On the other hand, the bipartite NLS of nucleoplasmin is comprised of two short stretches of basic amino acids that are separated by a flexible linker. Underscoring the evolutionary success of these transport signals, most of the nuclear proteins investigated so far contain one of the two types of NLSs (reviewed in Mattaj and Englmeier, 1998; reviewed in Nakielny and Dreyfuss, 1999; reviewed in Ström and Weis, 2001).

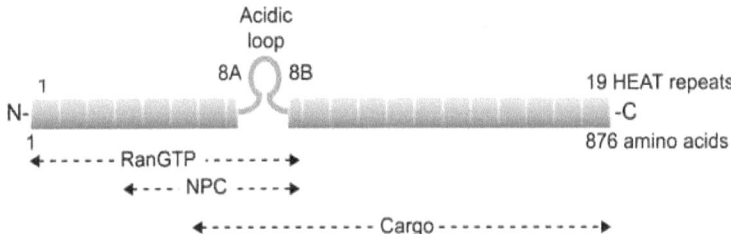

Figure 5: **Schematic structure of Importin β:**

Importin β contains 19 helical-repeat motifs (HEAT repeats) that are comprised of two helices connected by short turns. However, HEAT repeat 8 represents the only exception containing an acidic loop instead of the turn that is important for the regulation of cargo binding and release. The N-terminal HEAT repeats enable interaction of the receptor with the NPC but also RanGTP to release the substrate inside the nucleus. In contrast, cargo or importinα binding is managed by the repeats 7-19, partly overlapping the NPC and RanGTP interaction site. (Ström and Weis, 2001)

Export mediated by Exportin 5:

Similar to the importin α/β pathway, the best characterized export receptor, Exportin 1 or Crm1, binds to short peptide sequences that contain basic residues, such as leucines, isoleucines or valines, in a defined pattern (Wen et al., 1995; reviewed in Kutay and Güttinger, 2005). However, the Macara lab recently discovered a karyopherin, termed Exportin 5 (Exp5), that mediates export of dsRBD containing proteins by interacting with a dsRBD. Exp5 also belongs to the Impβ family and was initially identified as the export receptor transporting the interleukine enhancer binding factor 3 (ILF3, also known as NFAR1 or NF90) to the cytoplasm in a RanGTP dependent manner (Brownawell and Macara, 2002).

Interestingly, although inhibited by a 5' overhang, Exp5 shows high affinity to dsRNAs directly or bound by a dsRBD (Gwizdek et al., 2003). In fact, Exp5 has been found to shuttle specific classes of small RNAs to the cytoplasm. Although it was shown that the karyopherin binds and exports some tRNAs, it also facilitates export of micro-RNA precursors (pre-miRNAs) (Bohnsack et al., 2002; Calado et al., 2002; Yi et al., 2003; Lund et al., 2004). Pre-miRNAs are generated from large transcripts by the nuclear RNAse Drosha, producing about 60-70 nucleotide RNAs that form hairpins with short 3' overhangs (Lee et al., 2003). Exp5 specifically recognizes this type of structure and, subsequently, a trimeric complex with RanGTP moves out of the nucleus. In the cytoplasm the pre-miRNA is released and further processed into double-stranded 21bp-miRNAs by Dicer. Finally, one strand of the mature miRNA is then incorporated into the RISC complex and can fulfill its role in down-regulating a specific target gene (Hutvagner et al., 2001; reviewed in Bartel, 2004). Strikingly, it was recently revealed that overexpression of Exp5 enhances the mechanism of RNAi mediated by miRNAs but also the structural and functional similar short hairpin RNAs (shRNAs) (Yi et al., 2005).

As mentioned above, Exp5 was also found to export several dsRBD containing proteins in an RNA- and RanGTP dependent manner. For example, mammalian Staufen 2 is involved in mRNA localization in hippocampal neurons (Klieber et al., 1999). It is suggested that one of three known isoforms of the protein takes up its substrate RNA in the nucleus and is specifically exported via the Exp5 pathway. For that reason, one of the two dsRBDs serves as an Exp5 dependent NES, whereas single residues within this dsRBD and another dsRBD comprise an NLS (Macchi et al., 2004). Due to these findings, it is believed that Exp5 represents an alternative pathway to the well characterized heterogeneous nuclear ribonucleoprotein (hnRNP) dependent pathway, exporting mRNAs, in addition to its key role in localizing short, structured RNAs to the cytoplasm (reviewed in Lei and Silver, 2002; reviewed in Cullen, 2003).

Another interesting aspect of Exp5, underlining its outstanding range of functions and features, is that the yeast homologue of the karyopherin, Kap142p (or Msn5p), was shown to mediate import of, for example, replication protein A (Yoshida and Blobel, 2001), but also export of several other proteins (Komeili and O'Shea, 2001).

Moreover, recently the Exp5 homologue in *Arabidopsis thaliana*, HASTY, has also been enclosed to transfer miRNAs to the cytoplasm, similar to its human counterpart (Park et al., 2005).

Transportin 1 and a combined import and export signal: the M9 domain:

First discovered in hnRNP A1 that is involved in the metabolism and nuclear export of pre-mRNAs (Pinol-Roma and Dreyfuss, 1992; reviewed in Shyu and Wilkinson, 2000), the M9 domain facilitates bidirectional transport through the NPC. In contrast to other transport signals, the M9 domain is at least 38 amino acids in length and does not abundantly include basic residues (Siomi and Dreyfuss, 1995; Weighardt et al., 1995; Michael et al., 1995). M9 is recognized and bound by the Impβ related import receptor Transportin 1 (TRN-1, also termed karyopherin β2A) that moves hnRNP A1 efficiently across the nuclear envelope (Pollard et al., 1996; Bonifaci et al., 1997). Additionally, this domain also serves as an NES. Injection experiments of a mutated version of the transport signal in *Xenopus* oocytes resulted in a dramatic accumulation of mRNAs in the nucleus (Izaurralde et al., 1997). However, no karyopherin triggering export of hnRNP A1 has been detected so far.

Due to the fact that TRN-1 has also been found to interact with peptide sequences that share almost no homology to M9 in several other proteins, no general consensus sequence for TRN-1 binding could have been identified. For instance, basic lysine- and arginine-rich regions within the mRNA export factor TAP, specific for constitutive transport element (CTE) containing mRNAs (Gruter et al., 1998), and within the ribosomal proteins rpL23a, rpS7 and rpL5 are recognized by TRN-1. But in contrast to M9, these domains exhibit only NLS but not NES activity (Kang and Cullen, 1999; Bear et al., 1999; Truant et al., 1999; Bachi et al., 2000; Jäkel and Görlich, 1998). Furthermore, TRN-1 in addition to Transportin-2 (TRN-2) was found to cause nuclear accumulation of the mRNA-binding factor HuR that stabilizes mRNAs containing AU-rich sequences (AREs) within their 3'-UTR. Interestingly, HuR harbors a transport signal, named HNS, that partially matches the M9 domain (Fan and Steitz, 1998; Gallouzi and Steitz, 2001; Rebane et al., 2004; reviewed in Michael, 2000).

Nuclear transport of hsADAR1:

The biological importance of ADARs, as described above, resembles the necessity for ADAR activity to be highly regulated. Indeed, several mechanisms involved in the regulation of the protein have been suggested. These include alternative splicing, protein modification, protein sequestration in either compartment and, most prominently, regulated nucleo-cytoplasmic transport (Desterro et al., 2003; Liu et al., 1999; Palladino et al., 2000; Poulsen et al., 2001; Strehblow et al., 2002).

Previous experiments in our lab revealed that hsADAR1, in contrast to xlADAR1, is a transcription-dependent shuttling protein, constantly moving between the nucleus and the cytoplasm (Desterro et al., 2003; Eckmann et al., 2001, Sallacz and Jantsch, 2005). Furthermore, it was shown that the third dsRBD of hsADAR1 in addition to its ability to bind RNA acts as an atypical NLS. Strikingly, deletions on either end of this 70 amino acids long domain dramatically decreases NLS activity indicating that the 3D structure is required for an interaction with a karyopherin that carries the enzyme to the nucleus. Interestingly, RNA binding of this domain is not essential for nuclear import of the protein (Eckmann et al., 2001).

For nuclear exit, however, the longer version, hsADAR1i, but not the short form, hsADAR1c, contains a Crm1-dependent NES within the aminoterminal Z-DNA binding domain. This short, leucine-rich stretch of amino acids enables rapid escape from the nucleus to keep the nuclear substrate-ADAR1 ratio constant (Poulsen et al., 2001; Strehblow et al., 2002).

Consistent with the notion that the NES is missing from ADAR1c, we could demonstrate that at least one other region within hsADAR1i and hsADAR1c helps to regulate the enzyme concentration in the two compartments. On the one hand, dsRBD1 interferes with nuclear accumulation of a reporter construct containing dsRBD3 as an active NLS. On the other hand, some parts of the deaminase domain that could not have precisely been assessed also cause cytoplasmic accumulation of a reporter construct including the third dsRBD (Strehblow et al., 2002). However, the question whether the presence of dsRBD1 leads to interference of nuclear accumulation by preventing nuclear import or by stimulating nuclear export, has not been totally cleared and, hence, this region was termed Modulator of Import (MOI).

Interestingly, mutations that impair RNA-binding in either dsRBD1 or dsRBD3 fail to cause cytoplasmic accumulation. Furthermore, interference of nuclear accumulation was also obtained when the MOI was replaced by several other dsRBDs, for example xlADAR1 dsRBD1. However, among some other dsRBDs, the second dsRBD of xlRBPA, which had been found to be the best RNA binder *in vitro* (Krovat et al., 1996), failed to show this effect, indicating that the RNA binding strength is not crucial. In addition, cytoplasmic accumulation caused by the MOI is not transcription- dependent and is not transferable. Reporter constructs containing dsRBD1 and the NLS of the large T-antigen of SV40 could be clearly found in the nucleus.

A possible explanation includes binding of a common RNA by the two dsRBDs in the cytoplasm which leads to masking of the dsRBD-resident NLS (Strehblow et al., 2002). As mentioned before, a similar phenomenon has also been described for the *Xenopus* CBTF protein (Brzostowski et al., 2000), but has also recently been shown for the mammalian RNA helicase A (RHA, also known as DNA helicase II) that is involved in regulating gene expression at the transcriptional and post-transcriptional level (Aratani et al., 2001). In the latter case, the authors suggest that both dsRBDs within the enzyme contribute to anchor the protein in the cytoplasm. In addition, this process seems to be RNA-binding dependent (Fujita et al., 2005).

In contrast, some other dsRBDs have been found to interact with Exp5, as reported before. Although several findings regarding the MOI mentioned above rather indicate that the NLS-masking model is true, it cannot be excluded that dsRBD1 of hsADAR1 displays an RNA-binding dependent NES activity.

Despite the fact that ADAR1 of mice (mmADAR1) is extremely homologues to its human relative, its shuttling activity, however, has been found being regulated by at least three different regions within the protein that are distinct from those in hsADAR1. Although poorly shown, it should be mentioned that Nie and colleagues claim that nuclear import of mmADAR1 is mediated by an NLS in the C-terminus and export by an NES resident in the aminoterminus. Furthermore, a nucleolar localization signal (NoLS) that consists of a stretch of basic residues close to dsRBD3 might also keep the concentration of the protein in the nucleoplasm balanced by sequestering the enzyme to the nucleoli (Nie et al., 2004).

Specific aims of the thesis:

The main goal of this project has been to analyze the mechanism underlying the regulation of nuclear transport of hsADAR. Specifically, the role of dsRBDs in controlling this process has been investigated. Therefore, the following questions have been addressed:

1) Which elements within dsRBD3 of hsADAR1 are essential for NLS activity?

For that reason artificial dsRBDs were constructed which contained parts of hsADAR1 dsRBD3 fused to parts of other dsRBDs that show no nuclear accumulation when tested alone. These chimere dsRBDs in addition to multiple sequence alignments of dsRBDs with and without NLS activity and spatial considerations helped to identify candidate amino acids that were likely to interact with an import receptor.

Furthermore, a collaboration with the Allain lab in Switzerland has been started to analyze the structural singularity of this NLS-harboring dsRBD in comparison to other dsRBDs by NMR studies.

2) What is the import receptor of hsADAR1?

To address this questions several recombinantly expressed and purified import receptors have been tested in pull down and import assays using digitonin permeabilized tissue culture cells.

3) Is dsRBD1 of hsADAR1 an export signal, a cytoplasmic anchor, or does it act by masking the dsRBD3 resident NLS?

Since finding evidences for a cytoplasmic anchor or for NLS-masking is seemingly more difficult than studying export via binding of a karyopherin to the MOI, investigations were concentrated on the possible interaction of the most probable export receptor candidate Exp5 and dsRBD1. For that reason, several assays, including Co-immunoprecipitations and Exp5 RNAi, have been performed.

Although some results fail to support the idea of Exp5 mediating export of ADAR1, it can still not entirely be excluded and some more experiments have to be performed to clarify this question. However, if it turns out that Exp5 can export hsADAR1, it would be tempting to investigate if hsADAR1 also plays a role in transporting RNAs generally or a specific class of RNAs from the nucleus.

2. Materials and methods:

Cloning and plasmids:

Cloning of chimeric dsRBDs:

Two kinds of DNA primers were designed that enable PCR amplification of xlADAR1 dsRBD2 or xl4F.1 (the *Xenopus* homologue of the human NF90/NFAR-1) dsRBD1 and hsADAR1 dsRBD3 chimere dsRBDs. One set of primers was complementary to either end of the dsRBDs introducing an artificial *Hin*DIII restriction site. The second set of primers was chimeric; complementary at the 5' end to the region of the switching point of the two dsRBDs at the hsADAR1 dsRBD3 side and at the 3' end to the region of the switching point of the two dsRBDs at the xlADAR1 dsRBD2/xl4F.1 dsRBD1 side and vice versa. A first PCR reaction was set up, for example, with a *Hin*DIII primer complementary to the 5' end of hsADAR1 dsRBD3 and the chimeric primer. Using the hsADAR1 cDNA as template, PCR products could be obtained that contained the 5' hsADAR1 dsRBD3 part including an overhang complementary to the 5' end of the 3' xlADAR1 dsRBD2/xl4F.1 dsRBD1 part. This PCR product served as forward primer in a second PCR together with a reverse primer complementary to the 3' end of xlADAR1 dsRBD2/xl4F.1 dsRBD1 and xlADAR1/xl4F.1 cDNA as template. The resulting PCR product was then applied as template in a third PCR together with the formerly used *Hin*DIII primers. With this technique PCR products originated from two different DNA fragments were produced. Finally, the PCR products were purified, *Hin*DIII digested, and cloned into the previously published pyruvate kinase fusion construct vector (Eckmann et al., 2001; Strehblow et al., 2002).

In a similar way point-mutated xlADAR1 dsRBD2s were generated containing the full length consensus sequence of the dsRBD but having some codons switched to the corresponding codons from hsADAR1 dsRBD3. In a first PCR reaction, one of the *Hin*DIII primers complementary to a flanking sequence and a primer complementary to the site to be mutated, introducing nucleotides corresponding to the same region within hsDADR1 dsRBD3, produced a fragment that was used as a primer in a second reaction, together with the second *Hin*DIII primer. The resulting product was again used as template in a third PCR. The DNA from that reaction was subsequently separated on an agarose gel, the corresponding band was purified, *Hin*DIII digested and ligated into the pyruvate kinase fusion construct vector.

DsRBDs without any modifications were directly PCR-amplified from cDNA using only the flanking HinDIII primers. The DNA fragments were finally also ligated into the same vector.

This plasmid provides a start AUG that is upstream of a unique *Hin*DIII restriction site used for the insertion of the chimeric dsRBDs. Moreover, adjacent in frame downstream the pyruvate kinase and six tandemly arranged myc-epitopes are located. Positive clones were identified by restriction mapping and verified by sequencing.

Cloning of ADAR1c:

DNA molecules encoding ADAR1c were obtained by PCR from cDNA encoding full length ADAR1 by using primers complementary to either end of ADAR1c introducing a *Hin*DIII restriction site, as described before. This site was used to clone the DNA fragments into pEGFP-N2 or pEGFP-C2 (Clontech) providing a C-terminal or an N-terminal GFP-tag, into a CMV originated vector providing C-terminal 6xmyc-tags and a His-tag (a kind gift of A. Peters, Thomas Jenuwein Lab, Institute of Molecular Pathology (IMP), Vienna, Austria) and into the previously described pyruvate kinase fusion construct vector providing C-terminal 6x myc-tags.

Cloning of cDNA sequences for recombinant protein expression:

DNA fragments encoding the proteins of interest were cloned into an appropriate vector containing His tags, GST tags or were obtained from other labs. His-tagged proteins: Importin α and β in pRSET A, a kind gift of Emi Nagoshi, Osaka University, Osaka, Japan; Exportin 5 in pQE60, a kind gift of Ian G. Macara, University of Virginia, Charlottesville, VA; RanQ69L in pQE32, a kind gift of Ian Mattaj, EMBL, Heidelberg, Germany; mouse Importin 9b in pQE60, a kind gift of Dirk Görlich, ZMBH Heidelberg, Germany; Transportin-SR in pET-28a, a kind gift of Naoyuki Kataoka, Howard Hughes Medical Institute, University of Pennsylvania.
GST-tagged proteins: hsILF3 dsRBD2 in pGEX-2T, also a kind gift of Ian G. Macara; SV40-NLS-GFP in pGEX-2T, also a kind gift of Emi Nagoshi; all other GST-fusion constructs containing single or multiple dsRBDs were obtained by PCR amplification with hsADAR1 or xlADAR1 cDNA or with previously published PK-fusion constructs (Strehblow et al., 2002) as templates, including 6-10 amino acids of the N- or C-terminal flanking regions, and ligation into a *Hin*DIII

restriction site of a modified pGEX-1 vector containing an extended multiple cloning site which was available in our lab.

Expression and purification of recombinant proteins:

The vectors containing DNA sequences encoding for His- or GST-fusion proteins were transformed into the E. coli strain BL21 (DE3). The bacteria were grown in 100-600ml cultures containing the appropriate antibiotic (Ampicillin or Kanamycin) to an OD_{600} of 0,6-0,8 at 37°C. Protein expression was then induced by the addition of 0,75mM IPTG (isopropyl-βD-thiogalactopyranoside) and after another 3h incubation at 37°C in case of GST-fusion proteins or after incubation overnight at room temperature (RT) in case of His-tagged proteins the bacteria were harvested by centrifugation and lysed by sonication in an appropriate buffer (GST-proteins: saline phosphate buffer (PBS, 137mM NaCl, 2,7mM KCl, 4,3mM Na_2HPO_4, 1,4mM KH_2PO_4) pH 7,4, 1% Triton X100, 1mM PMSF; His-proteins: binding buffer: 20mM HEPES pH 7,3, 110mM KoAc, 2mM MgoAC, 20mM NaCl, 14mM β-Mercapto-ethanol, 0,5% bovine serum albumin (BSA), 0,1% Tween 20 and 1mM PMSF). After removing insoluble components by centrifugation for 10min at 10.000 rpm the recombinant proteins were purified using Ni-NTA agarose (Qiagen) or Glutathione-S-Transferase coupled agarose beads (Sigma) following the instructions of the producers. Briefly, about 0,5-1ml swollen and washed beads per 200ml culture were mixed with the bacteria lysates in a total volume of 10-20ml. After 1h incubation on a turntable at 4°C the mixture was transferred into a column. The beads were washed with approximately 20ml of the corresponding buffer and the bound proteins were eluted in 5-10 fractions 1ml each by pouring binding buffer containing 200mM Imidazol in case of His-tagged proteins or 50mM Tris buffer pH 9,5 containing 20mM Gluthatione for GST-proteins into the column.

Subsequently, the fractions were analyzed by Polyacrylamide gelelectrophoresis using sodium dodecylsulfate (SDS) (SDS-PAGE), denaturing the proteins, followed by Coomassie staining and fractions with abundant amounts of target protein were pooled and dialyzed against a buffer containing 20mM HEPES pH 7,3, 5% Glycerol and 7mM β-Mercapto-ethanol. Finally, the concentration of the purified proteins was measured using Bradford solution. The proteins were aliquoted and stored at -70°C.

Tissue culture:

Generally, HeLa and mouse NIH 3T3 cells were grown on 100 mm tissue culture plates in DMEM medium containing 10% Foetal calf serum (FCS) (both from PAA), 2mM L-Glutamine and 1% Penicillin/Streptomycin. For transfection assays, import assays or RNAi experiments cells were split on 20x20mm coverslips in 35mm tissue culture plates.

Chimere dsRBD constructs transfection and immunostaining:

Pyruvate kinase fusion constructs containing C-terminal myc tags were transfected into HeLa cells using the lipidbased transfection kit Metafectene (Biontex) or TFX20 (Promega), following the manufacturer's instructions. In principal, 500ng-2µg of DNA were mixed with 2-6µl of a transfection reagent in 300µl-1ml of DMEM medium without FCS and vigorously vortexed. After at least 20 min at RT the mixture was dropped on the cells. One additional hour later the plates were filled with medium containing FCS and left at least 24h at 37°C in an incubator providing 5% CO_2.

On the next day the cells were fixed with a solution containing 2% paraformaldehyde and permeabilized with cooled methanol on ice. After blocking with 10% horse serum (HS) in PBS pH 7,4, 0,05% Tween 20 (PBST), the cells were stained using a monoclonal antibody against myc-epitopes (9E10) in 2,5% HS in PBST in a 1:2 dilution and using an anti-mouse secondary antibody fluorescently labeled (Alexa 488), 1:500. Therefore, the localization of the expressed reporter proteins could be detected by fluorescence microscopy.

Actinomycin D treatment:

To inhibit transcription, Actinomycin D (AMD) (Sigma-Aldrich) was added to the tissue culture medium at final concentration of 40µg/ml 5-10h before harvesting of the cells.

Exportin-5 RNAi:

The sequences of both siExp5 RNA duplices used in this assay were already published and the synthetic siRNAs were obtained from MWG / Dharmacon Research Inc. One oligo was designed in the Cullen lab to target the nucleotides 339-360 of the human Exp5 ORF (Exp5 siRNA 1) (Yi et al., 2003), and the second was designed by Elsebeth Lund and colleagues (Exp5 siRNA 2: nt 160-179) (Lund et al., 2004).

Sequences of the Exp5 siRNAs:

Exp5 siRNA 1:

```
5'   NNGAUGCUCUGUCUCGAAUUGUA
3'   CUACGAGACAGAGCUUAACAUNN
```

Exp5 siRNA 2:

```
5'   NNGCCCUCAAGUUUUGUGAGG
3'   CGGGAGUUCAAAACACUCCNN
```

SiRNA transfections were performed with HeLa cells grown on coverslips at a final concentration of 50-100nM of each oligonucleotide-duplex using the lipidbased transfection kit Lipofectamine 2000 (Invitrogen) or Silentfect (BioRad) as recommended. Here, the reagents were similarly applied as explained for Metafectine and TFX-20. In some cases 24h after the first transfection the procedure was repeated and after 48-72h the cells were fixed, permeabilized, as described above, and stained using a monoclonal anti-Exp5 antibody (1:2000), a kind gift of Elsebeth Lund, University of Wisconsin Medical School, Madison, WI, to monitor Exp5 reduction *in vivo*. The knock down was further confirmed by western blotting using total HeLa extracts of siRNA transfected and untransfected cells. To visualize the proteins on the blots the same anti-Exp5 antibody and for control an anti-αActin antibody available in our lab was used. To detect the localization of transiently or stably expressed reporter proteins or endogenous hsADAR1 within the cells the previously noted anti-myc antibody or a monoclonal anti-hsADAR1 antibody (668) available in our lab were used.

Preparation of cytosolic, nuclear and total cell extracts:

To obtain cytosolic extracts, Hela cells were harvested and washed twice with PBS followed by lysis of the cells by 20min incubation on ice with lysis buffer (0,5mM Hepes pH 7,3, 0,75mM MgoAc, 0,15mM EGTA, 0,5mM PMSF, 3mM Dithiotreitol (DTT), Complete Mini protease inhibitor mix (Roche)) and by passing 10 times through a 0,45mm syringe needle. Thereby, cell breakage was checked using Trypan Blue staining (Sigma) and phase contrast microscopy. At this stage the cell membranes of the majority of the cells were burst, but the vast majority of nuclei have not been damaged. Subsequent 10min centrifugation at 10.000rpm separated cytosol from most of the insoluble components including nuclei. Further centrifugation at 35.000rpm removed any remaining insoluble particles.

For gaining nuclear extracts, the pellets derived from the first centrifugation step were resuspended in lysis buffer and passed at least another 20 times through the needle. Again, centrifugation removed insoluble components.

Preparation of total cell extracts was realized similar to nuclear extract except from centrifugation to obtain cytosol only. Ultimately, all cellular extracts were aliquoted and stored at -70°C.

Immunoprecipitations (IPs):

Interactions between individual dsRBDs and karyopherins were tested in pull down assays using extracts of HeLa cells or purified recombinant proteins.

IPs using cell extracts:

For precipitating HeLa-endogenous Exp5 by hsADAR1, monoclonal anti-hsADAR1 antibodies were coupled with protein A sepharose (Amersham Biosciences) and, subsequently, incubated with nuclear, cytoplasmic or total HeLa lysates in 500µl binding buffer at 4°C for 1h. After washing of the beads bound proteins were analyzed by western blotting using the previously mentioned anti-Exp5 antibody (1:4000).

IPs using recombinant proteins:

For testing interactions between recombinantly expressed and purified proteins, 500nM of GST- and 300nM of His-fusion proteins were incubated with approximately 20µl swollen GST beads (Sigma) at RT in a total volume of 500µl in binding buffer for 1h, as described above. The beads were three times washed with 1ml binding buffer and bound proteins were analyzed by SDS-PAGE and western blotting using monoclonal anti-His antibodies. For further detection secondary antibodies conjugated to alkaline phosphatase (AP) or horseradish peroxidase (HRP) were used (see below).

Gelelectrophoresis and western blotting:

Samples derived from pull down experiments were mixed with SDS-sample buffer and boiled in a water bath for 3min. Afterwards, the proteins were separated by 6-12% SDS-PAGE and transferred onto PVDF membranes (Osmonics) in a 25mM Tris buffer containing 200mM Glycine. After Ponceau S the blots were blocked with 5% milk in TBST and the proteins were detected by using appropriate primary antibodies.

In case of IP experiments, protein detection was performed by western blotting using a monoclonal anti-His antibody raised in rabbits (Rockland) 1:6000 or a monoclonal mouse-anti-His antibody (Qiagen), respectively. For further detection anti-rabbit or mouse antibodies conjugated to alkaline phosphatase (1:2000) or horse radish peroxidase (1:20000) (both Sigma) were used.

Import assays:

HeLa or NIH 3T3 cells grown on coverslips were washed with import buffer (20mM HEPES pH 7,3, 110mM KoAc, 2mM MgoAC, 1mM EGTA, 1mM DTT and 1mM PMSF) and treated with digitonin in final concentration of 30µg/ml in ice-cold import buffer for 5min to permeabilize the cell membrane. Thereby, the permeabilization status of the cells was monitored by Trypan Blue staining. After three times washing of the cells removing the endogenous cytoplasm, the cells were incubated with a mix of a recombinantly expressed putative NLS containing proteins tagged with GST, cytosolic HeLa extracts or purified karyopherins and an energy source (see below). The import machinery present in the cytosolic fraction or the karyopherins had translocated the protein to the nucleus if an NLS recognizable by the import factors was present.

A typical 50µl reaction mix contained:
- 1µM NLS-protein,
- 1µM of a karyopherin or 25µl cytosol (4mg protein/ml),
- 5µl 10x import buffer,
- 20U/ml creatine phosphokinase,
- 5mM creatine phosphate,
- 1mM rGTP, rATP,
- 0,1% BSA,
- 1mM PMSF

Subsequently after 30min incubation at 30°C in a moist chamber the cells on the coverslips were fixed as described above. Staining of the GST-proteins was performed by using a monoclonal goat-αGST-antibody (Rockland) 1:500, followed by an αgoat antibody fluorescently labeled (Alexa 488), 1:500. Again, detection was executed by fluorescence microscopy.

Affinity chromatography and mass spectrometry:

To find interacting partners of ADAR1 dsRBD3, especially focussing on karyopherins, approximately 1,5mg recombinantly expressed and purified GST-dsRBD3 was loaded onto a 1ml HiTrap NHS-activated column (Amersham Pharmacia) in coupling buffer (0,2M NaHCO3, 0,5M NaCl, pH 8,3) over night at 4°C. Deactivation and sealing of the still unoccupied residues was performed by washing the column thrice with 6ml 0,5M ethanolamine, 0,5M NaCl, pH 8,3, followed by 6ml 0,1M acetate, 0,5M NaCl, pH 4. After 30min incubation at RT the column was finally washed with 6ml coupling buffer to neutralize the deactivation solutions and thereby improving correct folding of the covalently bound proteins.

The actual assay was carried out by circulation of 5-10ml cytosolic or nuclear extracts of HeLa cells in import buffer (0,5-1mg/ml total protein concentration) through the column at RT for 30min. The column was subsequently washed with 10ml import buffer to remove unbound proteins. Elution of ADAR1 dsRBD3 interacting partners took place in three steps. The first fraction was obtained by shortly incubating the column with 1ml recombinantly expressed and purified His-RanGTP in import buffer (2mg/ml protein concentration) to terminate any import receptor-NLS specific interactions. The second and third elution step was performed by washing the column with 1ml 0,5M and 1M NaCl, respectively. To remove RanGTP from the first fraction what would inhibit nuclear import when tested in import assays, the fraction was incubated with 1ml Ni-agarose in import buffer at 4°C for 1h. Similar, the second and third fractions were dialyzed against import buffer at 4°C for 4h to normalize the NaCl concentrations to 20mM.

The eluted proteins in the fractions were analyzed by separation on a 7,5-17% SDS-PAGE gradientgel and visualized by silver staining: Initially, the proteins were fixed on the gel to avoid further diffusion by incubating the gel 20min in 100ml fixer solution (50% methanol, 5% acidic acid). To reduce background the gel was then washed twice. First, the gel was washed 10min in 100ml washing solution (50% methanol) and then in 100ml water over night. After a 1min incubation period of the gel in 100ml sensitizing solution (0,02% sodium thiophosphate ($Na_2S_2O_3$)) the gel was washed again twice, but this time only 1min in water. Then, the gel was exposed to a 100ml silver nitrate solution (0,1% AgNO3) at 4°C for 20min. Again the gel was washed twice in water for 1min, before the gel was developed by incubation in 100ml developer solution (2% Na_2CO_3, 35% formaldehyde) for 0,5 to 5min. After the bands were visible, development was terminated by washing the gel three times with stop solution (5% acidic acid) for 1min.

Finally, protein bands in the mass range of 60-130kDa were cut out and subjected to MALDI-TOF mass spectrometric analysis in the Vienna Biocenter Mass Spectrometry Unit.

3. Results:

Nuclear Import of hsADAR1:

Structural analysis of dsRBD3 as an active NLS:

Identification of candidate amino acids:

Although dsRBDs have been previously known to fulfill additional functions, except from RNA binding, including protein-protein-interactions, hsADAR1 dsRBD3 was the first dsRBD identified acting as a nuclear transport signal. Despite the fact that the vast majority of all NLSs are comprised of short, basic amino acid sequences, this import signal spans a 70 residue, highly structured domain. In fact, the obvious presumption that correct folding of the complex 3D structure of the dsRBD ensures that NLS-relevant residues are located at the right spatial position within the dsRBD enabling interaction with a karyopherin, is supported by the finding that deletions from either end dramatically reduce import efficiency in transfection based reporter construct assays. But although it is known that the third dsRBD of hsADAR1 is essential for substrate recognition and RNA binding of the enzyme, mutations impairing these functions do not harm NLS activity (Eckmann et al., 2001).

Given the fact that dsRBDs are generally very conserved, the question has been addressed which amino acid residues constitute this atypical NLS that are not included in other dsRBDs. In an initial step a few dsRBDs from different proteins were tested for potential NLS activity. These included xlADAR1 dsRBD2 and 3 (*Fig. 7: xlNLS18 and 17*), dsRBD1 and 2 of xl4F.1, which is a homologue of the human NF90/NFAR-1 protein, involved in mRNA processing, (NLS192 and NLS193, data not shown) and dsRBD2 of *Xenopus laevis* RNA binding protein A (xlRBPA) (NLS194, data not shown). The domains were tested for their ability to enter the nucleus in the previously described transfection based reporter construct assay (Eckmann et al., 2001). Interestingly, several dsRBDs failed to accumulate in the nucleus, except from the third dsRBD of xlADAR1. Apparently, the frog version of this dsRBD is so homologous to its human counterpart that it can be recognized by the same import receptor in HeLa cells, even though it does not display NLS activity in *Xenopus* oocytes (Eckmann et al., 2001). However, it should be mentioned that in some cells expressing the second dsRBD of xlADAR1 weak nuclear signals could be observed indicating that an import receptor exhibits at least some affinity to this dsRBD when overexpressed in HeLa cells.

The finding that dsRBD3 of the amphibian ADAR1 also accumulates in the nucleus of HeLas was further used to identify residues that can be found in the two NLS-active dsRBDs, but are excluded from any other dsRBD tested. For that reason, multiple amino acid sequence alignments were generated enabling convenient finding of identical residues in the two dsRBD3 versions as potential NLS-comprising amino acids, but that are different in the four other RNA binding domains (*Fig.6*). Moreover, the remaining 28 candidate residues were checked whether they consistently occur in several other dsRBDs of other proteins that were not tested. Conserved amino acids had to be eliminated as candidates because they are obviously responsible for RNA binding or are important for building up the structure of the domain and would be therefore not in charge of import factor interactions. Finally, 18 potential NLS amino acids could be identified.

Figure 6: **Multiple dsRBD alignment:**
Comparison of the third dsRBDs of hsADAR1 and xlADAR1 (highlighted in green) with xlADAR1 dsRBD2, xl4F.1 dsRBD1, 2 and xlRBPA dsRBD2, which had all been negatively tested for NLS activity. Only amino acids that are identical in xlADAR1 and hsADAR1 dsRBD3 but not in any other of the mentioned dsRBDs should build up the dsRBD3-resident NLS. Candidate amino acids essential for NLS activity that are exclusively located in these two dsRBDs are highlighted in red. In fact, 28 amino acids within dsRBD3 of hsADAR1 fulfill this criterion. Interestingly, all N-terminal extensions show no homology

to each other while the C-terminal extensions of xlADAR1 dsRBD3 and hsADAR1 dsRBD3 are very conserved. Identical amino acids are shown in yellow, similar residues are boxed.

The next step to search for NLS-relevant residues was to construct chimeric dsRBDs, which should help to narrow important amino acids to a certain region within the dsRBD. These artificial domains consisted of different parts of hsADAR1 dsRBD3 and xlADAR1 dsRBD2 or xl4F.1 dsRBD1 fused at exactly the same position ensuring that the structure of the chimeric dsRBDs was maintained. Consequently, these domains contained amino acids from the human dsRBD, which could potentially comprise the atypical NLS, and amino acids from the amphibian dsRBDs. Thus, chimeras containing all residues essential for NLS activity should be able to enter the nucleus.

Six chimeric dsRBDs were constructed and tested for NLS activity. (Construct NLS138: hsADAR1 dsRBD3 708-743 + xlADAR1 dsRBD2 697-759; NLS139: xlADAR1 dsRBD2 664-696 + hsADAR1 dsRBD3 744-801; NLS188: xlADAR1 dsRBD2 677-749 + hsADAR1 dsRBD3 708-722 + 796-801; NLS189: xlADAR1 dsRBD2 664-676 + 750-759 + hsADAR1 dsRBD3 722-795; NLS198: hsADAR1 dsRBD3 708-743 + xl4F.1 dsRBD1 418-482; NLS199: xl4F.1 dsRBD1 394-417+ hsADAR1 dsRBD3 744-801)

Surprisingly, all of these dsRBDs accumulated in the cytoplasm strongly indicating that the NLS-relevant residues are spread over the entire dsRBD3 of hsADAR1 and are not concentrated in a single cluster (*Fig. 7*). Furthermore, even a construct containing the complete NLS-resident dsRBD flanked by at least 10 amino acids of xlADAR1 dsRBD2 could not be imported (NLS189, data not shown). Therefore, also residues besides the dsRBD consensus sequence had to be considered as candidates. Consistently, two essential residues could have been identified by accident. A construct containing xlADAR1 dsRBD3 with an artificially introduced KpnI restriction site in the 3' flanking region converting a lysine and an alanine residue to a glycine and a thyrosine residue failed to enter the nucleus in contrast to a wild type version of the dsRBD (xlNLS16: xlADAR1 dsRBD3 aa K868A, G869T, data not shown).

Assuming that all residues displaying a karyopherin interaction should be located on the surface of a dsRBD and should not be covered when RNA is bound, further candidates obtained from the dsRBD sequence comparisons were excluded. Using crystal structure data of the second dsRBD of xlRBPA (Ryter and Schultz, 1998) and the computer software Cn3D 4.1, the relative spatial positions of the amino acids identified above were determined. Interestingly, the remaining candidates built up a cluster, which is located at the opposite side of the RNA interaction site while in the alignment these residues are distributed over the entire dsRBD (*Fig.8 A, B*). In the end, 11 potential NLS-comprising amino acids could be identified (*Fig.8 C*).

xlADAR1
hsADAR1
FITC
DAPI
NOM
NLS35
xlNLS17
xlNLS18
NLS138
NLS139
NLS188

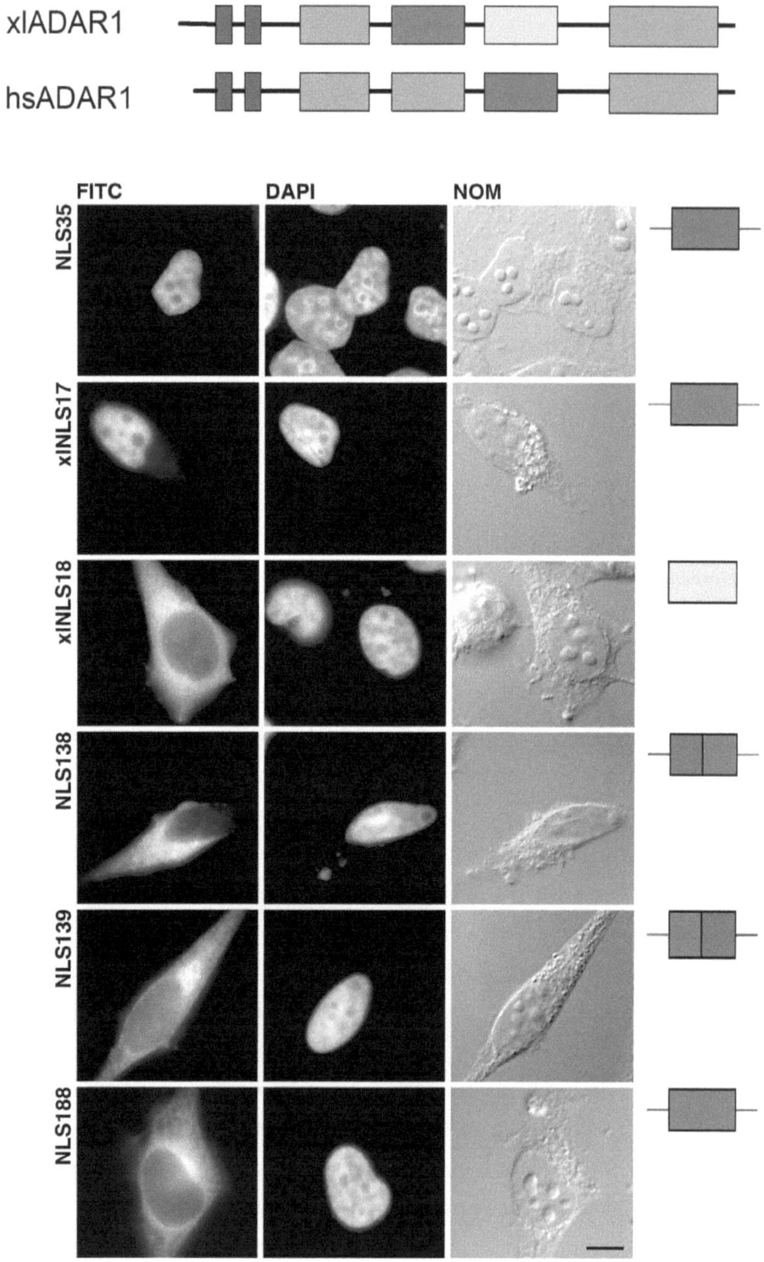

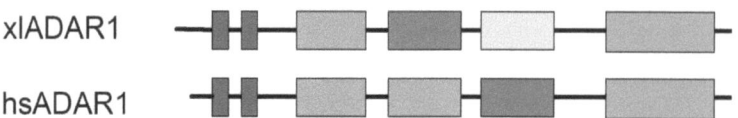

Figure 7: **Residues essential for NLS activity are distributed throughout the entire dsRBD3 of hsADAR1.**
Functional domains of hsADAR1 and xlADAR1 are illustrated with colored boxes connected by lines. While constructs containing dsRBD3 of hsADAR1 (NLS35, blue box) and xlADAR1 (xlNLS17, yellow box) can enter the nucleus of HeLa cells, the second dsRBD of xlADAR1 (xlNLS18, red box) and other dsRBDs tested cannot. Chimeric dsRBD constructs combining two parts different dsRBDs at exactly the same position maintaining the 3D structure obviously fail to interact with an import receptor, indicating that NLS-comprising residues are spread over the entire dsRBD3. Therefore, construct NLS138, containing the human N-terminus of dsRBD3 and the Xenopus C-terminus of dsRBD2 of ADAR1 (blue/red box), accumulates in the cytoplasm, as well as NLS139, harboring exactly the opposite parts of the two dsRBDs (red/blue box), and NLS188, containing the entire consensus sequence of xlADAR1 dsRBD2 and the flanking regions of hsADAR1 dsRBD3 (red box with blue lines). Deaminase domains are indicated by gray boxes, dsRBDs1 and 2 of xlADAR1 and dsRBD1 of hsADAR1 by green boxes and ZBDs by brown boxes.

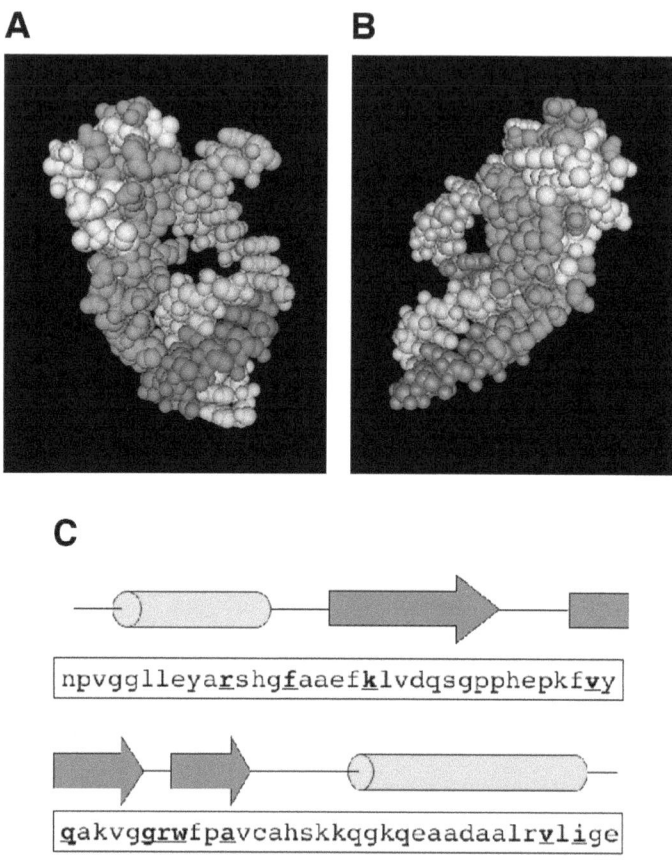

Figure 8: **Identification of surface exposed amino acids conserved in NLS bearing dsRBDs:**
(A+B) To reduce the number of candidate amino acids essential for NLS activity, surface-exposed amino acids were identified. Using the crystal structure of the second dsRBD of xlRBPA (violet and yellow) with bound dsRNA (gray, brown and green), here shown from two different angles (Ryter and Schultz, 1998), the spatial positions of the amino acids identified above were determined. Assuming that all residues displaying a karyopherin interaction should be located on the surface of a dsRBD and should not be covered when RNA is bound, further candidates were excluded. Interestingly, the remaining potential NLS-comprising amino acids (yellow) build up a cluster, which is located at the opposite side of the RNA interaction site while in the alignment these residues are distributed over the entire dsRBD. (C) The complex 3D structure of dsRBDs is comprised of two α-helices (green cylinders) enclosing three β-sheets (brown arrows). The 11 identified

NLS-candidate amino acids within hsADAR1 dsRBD3 (bold and underlined) are distributed throughout the entire dsRBD regardless in which structural region they were located.

Construction of mutated dsRBDs:

To examine whether the 11 candidate residues are the actual NLS-comprising amino acids, capable of contacting an import receptor, the corresponding residues within dsRBD2 of xlADAR1 were step wisely switched into that of the NLS-dsRBD. If all essential amino acids were introduced in such a replacement construct, it would be able to accumulate in the nucleus. NLS188, containing the second dsRBD of xlADAR1 was used as a starting point. Flanking to this dsRBD the regions surrounding dsRBD3 of hsADAR1 were inserted, since we could show that this flanking regions are also required for proper nuclear transport. Initially, four constructs were made with two to four single mutations (NLS202: Q690R + N694F + Q699K; NLS203: T713V + T715Q; NLS204: N720G + Q721R + T722W + P725A; NLS205: E745V + L747I). Presumably due to the fact that the mutations on every single construct were rather clustered and not distributed throughout the dsRBD, none of the four constructs could be imported.

Therefore, a construct was made having all 11 candidate amino acids in the non-NLS dsRBD converted into that from dsRBD3 of hsADAR1 and tested for its ability to be transported to the nucleus. Most surprisingly, however, even this dsRBD-construct accumulated in the cytoplasm.

In summary, this outcome either suggests that not all NLS-comprising residues have been considered as candidates and also conserved amino acids might be important, or that hsADAR1 dsRBD3 folds into a slightly different 3D structure compared to other dsRBDs exposing the NLS-residues in a unique way.

Ongoing experiments:

With the help of NMR analysis the question whether the third dsRBD of hsADAR1 has a 3D structure different from other dsRBDs will be addressed. Furthermore, exposed residues will be identified directly on this dsRBD that potentially comprise an NLS. For that reason a collaboration with Frederic Allain and his colleagues at the ETH Zürich, Switzerland, has been started. However, first attempts to express and purify a His-tagged version of the NLS-bearing dsRBD failed due to insolubility of the protein. But, alternative strategies and new efforts are undertaken to obtain sufficient amounts of dsRBD3 for NMR studies.

Interaction of import receptors with dsRBD3 *in vitro*:

Considering the fact the third dsRBD of hsADAR1 is an atypical NLS hidden in a complex protein structure, it is not apparent what import receptor can interact with this domain and can trigger transport of the enzyme into the nucleus. To find out which karyopherin can bind to the dsRBD, several import factors that seemed to be probable candidates were tested in immunoprecipation assays (IPs). In fact, recombinantly expressed and purified Importinα, β, Importin 9, a 130kDa homologue of Importin β that was found to mediate nuclear import of ribosomal proteins via binding to structured, basic domains (Jäkel et al., 2002), and Transportin-SR, which interacts with proteins containing serine-arginine-rich domains, were checked in pull down assays for their ability to bind dsRBD3 (*Fig. 9*). For negative control ADAR1 dsRBD2, exhibiting no NLS activity (Eckmann et al., 2001) and for positive control a GFP tagged SV40-NLS in case of Impα was used. Due to lack of substrates for Impβ, TRN-SR and Imp9, the experiment had to be performed without positive controls for these karyopherins.

Surprisingly, only Impα showed stronger affinity to the dsRBD-resident NLS than to the negative controls. Despite the fact that in the IPs with Impα high background can be observed, the import receptor binds dsRBD3 as strongly as the positive control SV40 NLS (*Fig.9 A*). In contrast, Impβ could bind ADAR1's NLS as efficient as the two negative controls (*Fig.9 B*). Interestingly, in case of Imp9 the IPs with hsADAR1 dsRBD2 and SV40 NLS generated the strongest signals what seemingly results from unspecific binding (*Fig.9 C*). However, TRN-SR showed no binding to dsRBD3 at all but weakly to the NLS of SV40 (*Fig.9 D*).

Given the fact that Impα usually binds to classical, basic NLSs, the results of the IPs indicate that neither Impα nor the other tested karyopherins can interact with ADAR1 dsRBD3 specifically. Furthermore, as described in the introduction, Impα serves exclusively as an adaptor protein for Impβ implying that binding of this import receptor to dsRBD3 should be enhanced when incubated together with Impα. But, in this case no signal could be detected at all (*Fig.9 D*). Therefore, these results suggest that in this experiment a general affinity of import receptors to basic sequences could be observed. However, it cannot be excluded that Impα takes part in nuclear import of hsADAR1.

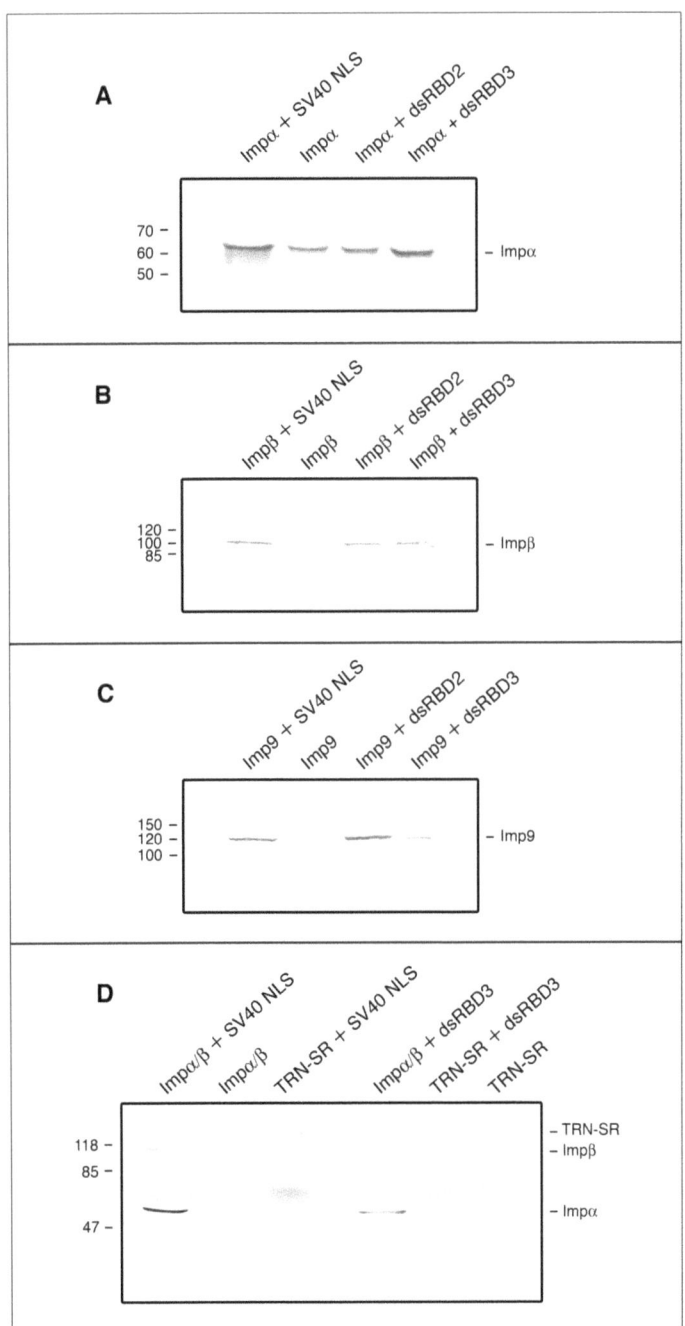

Figure 9: **Pull down assays with dsRBD3 and several import receptors:**
(A) Importinα binds to hsADAR1 dsRBD3 as efficient as to the positive control, the NLS of the large T-antigen of SV40, while an IP with Impα and hsADAR1 dsRBD2 shows no interaction of the two proteins. The corresponding band is as intense as the signal of the negative control, His-tagged Impα incubated with GST-beads, which produces a high background. (B) In contrast, Impβ exhibits almost no affinity to the beads. The karyopherin can bind to dsRBD2 and 3, but also to the SV40 NLS which was used as a negative control, indicating unspecific interactions. (C) Interestingly, importin 9 binds strongest to the second dsRBD of hsADAR1 but weakly to the third. Additionally, the importin shows some affinity to the SV40 NLS. (D) In contrast to the positive control (Impα/β + SV40 NLS) where a weak Impβ signal can be seen resulting from binding of this import receptor to its adaptor, dsRBD3 can only be bound by Impα. Transportin-SR cannot be detected when incubated with dsRBD3 but in an IP with the NLS of SV40. (Protein sizes in kDa)

DsRBD3 of hsADAR1 can be imported by several karyopherins in import assays:

To analyze if the import receptors tested in the IPs can transport ADAR1's NLS into the nucleus *in situ,* import assays were performed. For that reason digitonin permeabilized tissue culture cells without endogenous cytosol but intact nucleus were incubated with recombinantly expressed dsRBD3, an import receptor, and an energy source. In principle, only specific binding of the karyopherin to the NLS should enable the dsRBD to accumulate in the nucleus. Enrichment in nucleoli was also expected since dsRBDs generally exhibit high affinity to structured dsRNA that is highly enriched in these compartments. In addition, when overexpressed in tissue culture cells, nucleolar accumulation of dsRBDs can also be perceived. Consistently, nucleolar sequestration of ADAR2 has recently been proposed as an alternative mechanism to modulate the enzyme's activity (Sansam et al., 2003).

As depicted in *Fig.10*, the negative control, dsRBD3 without further proteins in the reaction mix, was not capable of entering the nucleus (*Fig.10: dsRBD3*). However, in some cells little nuclear staining could be observed, what seemingly resulted from incomplete removal of endogenous cytoplasm. In contrast, dsRBD3 incubated with cytosolic extracts was transported into nucleus (*Fig.10: dsRBD3 + cyt*). Even though a proportion of the added dsRBD3 remained in the cytoplasm, it can be stated that the NLS is functional and the corresponding import receptor was contained in the cytosol. The finding that import was not complete in this assay can be explained by the fact that the concentrations of karyopherins in the added cytosol are relatively low and, hence, not sufficient to import every protein molecule within a cell. On the other hand, dsRBDs generally tend to multimerize and agglutinate especially when bound to RNA, thereby masking the NLS.

Consequently, complete import of dsRBD containing proteins can hardly be achieved performing this assay. Nevertheless, by using a fusion protein harboring the classical NLS of xlADAR1 and *Xenopus* cytosol in mouse 3T3 cells total nuclear accumulation could be observed (*Fig.10: xlNLS + xlCyt*).

To investigate, whether the import assays imitate the *in vivo* situation, a protein containing all three dsRBDs of hsADAR1 was tested for import in the presence of cytosol. In agreement with transfection assays the fusion protein was found in the cytoplasmic compartment underlining the reliability of this assay (*Fig.10: dsRBD1-3 + cyt*).

As mentioned above, the same four karyopherins previously applied in IPs were also tested in import assays for their ability to mediate import of dsRBD3. As can be seen in *Fig.10*, using 1µM import receptor, every single karyopherin showed at least weak affinity to the NLS and could transport a certain proportion of the protein into the nucleus of HeLa cells. In case of Impα, which solely applied should not be able to transport dsRBD3 into the nucleus, it can be speculated that the presence of this import factor enhances the background by providing sufficient amounts of the adaptor protein for residual endogenous Impβ. However, a combination of both Impα and Impβ was found to import the NLS as efficiently as Impβ alone (*Fig.10: dsRBD3 + Impα; dsRBD3 + Impβ; dsRBD3 + Impα/β*). Unfortunately, the positive control, a fusion protein containing the NLS of SV40, incubated with Impα/β did not accumulate in the nucleus (data not shown).

As a consequence, these findings rather suggest that all karyopherins tested trigger unspecific import of dsRBD3 of hsADAR1 in import assays using HeLa cells. This statement can be supported by the fact that in mouse 3T3 cells almost no nuclear accumulation could be observed by using these purified karyopherins, while by using HeLa cytosol the NLS was detected in the nucleus (data not shown).

Again, this result can be explained by the fact that in abundance every import receptor displays a certain affinity to basic, positively charged domains. For example, the ribosomal protein S7 (rpS7) can be transported into the nucleus of HeLa cells by several karyopherins, including Transportin-1, Impα/β and Imp9, performing import assays when sufficient amounts of importins are present (Jäkel and Görlich, 1998; Jäkel et al., 2002).

Another interesting observation was made when import assays were performed with dsRBD3 and RanGTP alone without addition of other proteins. Theoretically, the dsRBD should remain in the cytoplasm because no karyopherin is present in the cytoplasmic compartment. Even if there were not sufficiently removed endogenous import receptors, the fusion protein would stay outside the nucleus because RanGTP should destroy any import factor-NLS interaction.

To our surprise, however, the NLS-bearing dsRBD was efficiently imported and accumulated in nucleoli (*Fig. 10: dsRBD3 + RanGTP*). To investigate this unexpected effect we determined the localization of the recombinantly expressed GTP associated GTPase within the permeabilized cells. Interestingly, it turned out that RanGTP accumulates immediately in the nucleus and would therefore not harm any NLS-karyopherin dimer in the cytoplasm. For that reason it is tempting to speculate that the dsRBD could be imported because RanGTP triggered release of the corresponding import receptor from the nucleus. However, no evidence of that mechanism could have been found and, thus, it remains unclear how the NLS can be imported in the presence of RanGTP.

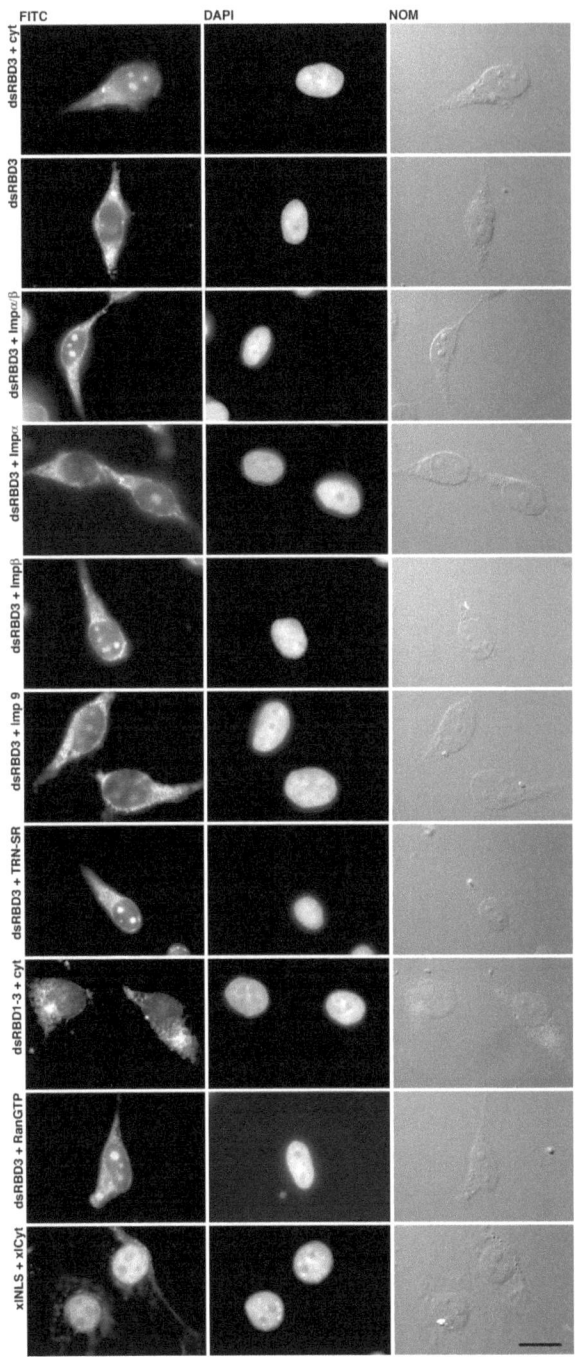

Figure 10: **Several karyopherins facilitate nuclear import of the dsRBD3-resident NLS in import assays:**

Import factors in cytosolic extracts trigger nuclear accumulation of ADAR1's NLS where the dsRBD associates with nucleoli (dsRBD3 + cyt). In contrast, in digitonin permeabilized HeLa cells incubated with dsRBD3 without any exogenously added karyopherin, the NLS is excluded from the nucleus. Consistent with transfection assays, a protein containing all three dsRBDs of hsADAR1 was found almost exclusively in the cytoplasmic compartment even when incubated with cytosolic extracts (dsRBD1-3 + cyt). In assays set up with 1µM of various karyopherins, total nuclear accumulation of dsRBD3 was not observed but every import receptor tested generated nuclear signals. Especially TRN-SR and a combination of Impα and Impβ mediated import of hsADAR1 dsRBD3 quite efficiently (dsRBD3 + TRN-SR, dsRBD3 + Impα/β), compared to Imp9 or Impα alone (dsRBD3 + Imp9, dsRBD3 + Impα). However, for unsolved reasons also RanGTP triggered nuclear uptake of the NLS (dsRBD3 + RanGTP). Almost complete nuclear accumulation of an NLS-protein could only be achieved by adding a protein containing classical NLS of xlADAR1 and Xenopus cytosol to permeabilized mouse 3T3 cells (xlNLS + xlCyt). Bar = 20µm.

Transportin-1: a possible candidate:

An alternative approach to identify a karyopherin responsible for nuclear import of hsADAR1 was pursued by performing dsRBD3-affinity chromatography followed by mass spectrometric analysis. Recombinantly expressed dsRBD3 was immobilized in a column and exposed to total cytosolic extracts of HeLa cells under appropriate conditions. The idea was that import factors and proteins with high affinity to the dsRBD should be kept within the column, whereas other proteins in the cytosol should be washed out. To elute specifically bound karyopherins, in a first step the column was washed with RanGTP. As described in the introduction, this GTPase in the GTP associated form triggers release of cargo proteins from their import receptors in the nucleus and, thus, it should remove any import factor bound to the dsRBD3-resident NLS in the column. All remaining proteins were ultimately eluted by washing the column with increasing NaCl concentrations and collected in fractions. After separation of the proteins, four bands in the mass range of 60-130kDa were cut out and their components were analyzed by mass spectrometry (MALDI-TOF) (*Fig.11 B*).

Strikingly, amongst several other proteins not participating in nucleocytoplasmic transport, Transportin-1 could be identified in one of the bands with the estimated size of 85kDa. As mentioned before, TRN-1 interacts with the previously described atypical transport signal, the M9 domain, which facilitates bidirectional transport through the NPC. Considering that there exists no consensus sequence in the import signals of its various substrates and that TRN-1 mediates import of several RNA- and DNA-binding proteins, the karyopherin is formerly a good candidate for importing hsADAR1.

To test whether TRN-1 is capable of specifically transporting dsRBD3 into the nucleus, import assays with the fraction obtained from affinity chromatography were performed. For comparison the previously tested import receptors were included in the experiment. To use approximately equal amounts of karyopherins the intensities of recombinantly purified import receptors and affinity purified TRN-1 were compared on a Coommassie stained gel. In fact, 38nM of each import receptor, corresponding to about 1/25th of the karyopherin concentration used in previous import assays, were added to each experiment.

As depicted in *Fig.11 A*, the fraction containing TRN-1 transported a proportion of dsRBD3 into the nucleus, whereas the four recombinantly expressed karyopherins were not able to trigger nuclear accumulation of the NLS. Furthermore, even a 1:10 dilution of the protein fraction produced some nuclear signals. Assuming that TRN-1 had only comprised a fraction of the band that was analyzed by mass spectrometry, one can speculate that TRN-1 was at least 20 times more efficient in the import assays what implies specificity for that NLS.

To further test the involvement of TRN-1 in nuclear import of ADAR1, recombinantly expressed and purified TRN-1 will be tested for its ability to mediate nuclear import of the dsRBD3-resident NLS in near future.

Taking together, TRN-1 seems prime candidate import receptor of hsADAR1 performing transport of the enzyme to the nucleus of living cells.

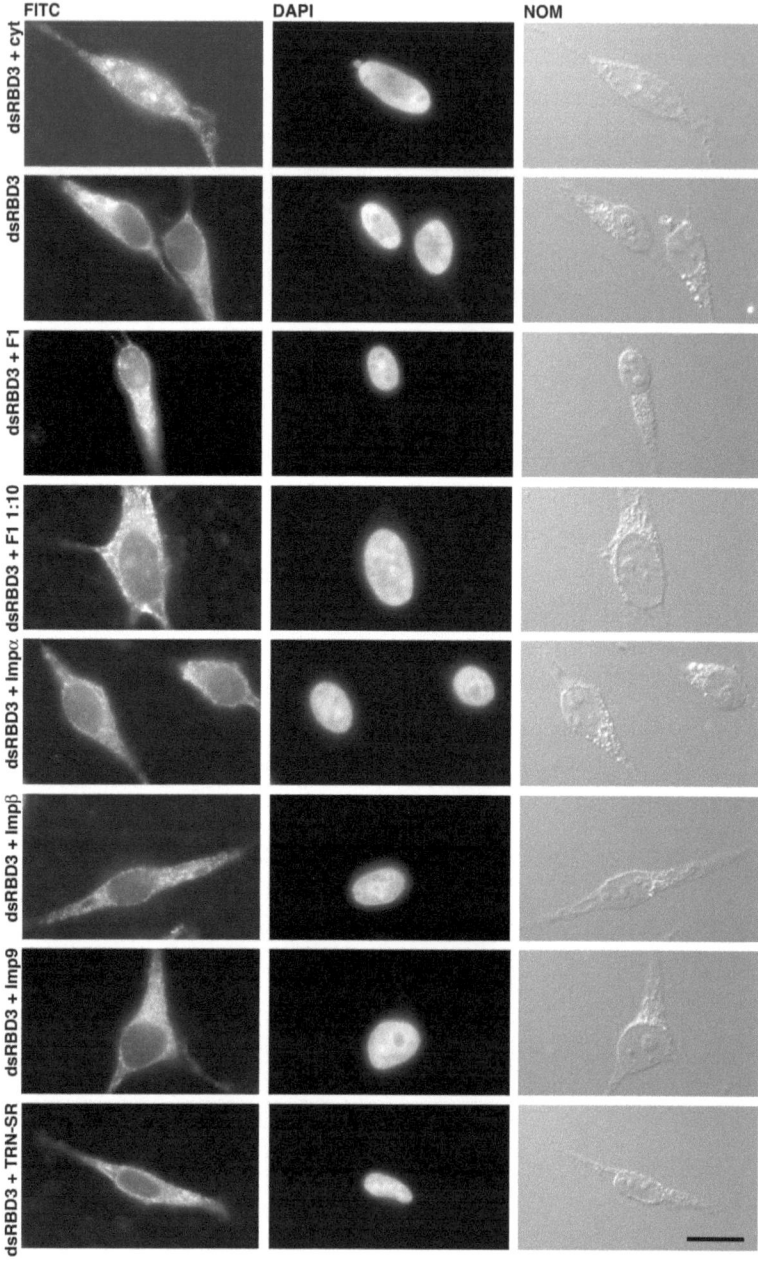

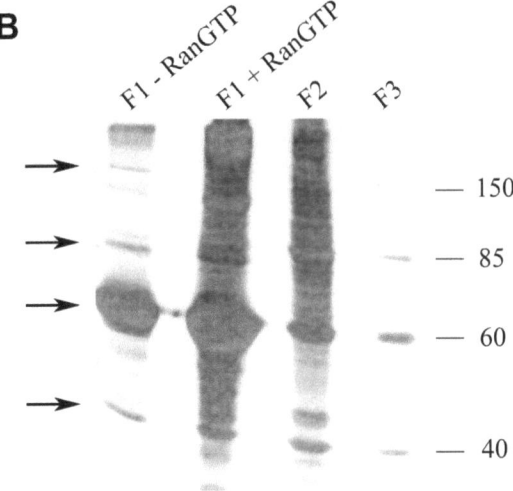

Figure 11: **DsRBD3 can be imported by a TRN-1 containing affinity chromatography fraction but not by several other recombinantly expressed importins using equal concentrations**

(A) HsADAR1 dsRBD3 can enter the nucleus of digitonin permeabilized HeLa cells where it accumulates in the nucleoli in the presence of cytosolic extracts, while the NLS remains in the cytoplasm in the absence of cytosol (dsRBD3 + cyt, dsRBD3). In contrast to several import receptors (dsRBD3 + Impα, + Impβ, + Imp9, + TRN-SR), a protein fraction containing TRN-1 obtained from a dsRBD3-affinity column that had been incubated with cytosol and the bound proteins had been eluted with RanGTP, caused nuclear/nucleolar accumulation of a small proportion of the NLS-harboring dsRBD (dsRBD3 + F1). Even a 1:10 dilution of the fraction was sufficient to mediate import of some fusion proteins, indicating a specific interaction of the karyopherin and the RNA-binding domain (dsRBD3 + F1 1:10), even though the majority of the NLS-containing protein was localized in the cytoplasmic compartment. Bar = 20µm.

(B) Four bands (indicated by arrows) were cut out of the first (left) lane of a silverstained gradient gel and subjected to mass spectrometry. In the first two lanes the first fraction of the dsRBD3 affinity chromatography, which had been previously eluted with RanGTP, was loaded. Left: RanGTP depleted fraction 1 (F1 – RanGTP); right: F1 still containing RanGTP (F1 + RanGTP). In the third and fourth lane the second and third fraction were loaded, which had been eluted with 0,5 and 1M NaCl, respectively (F2, F3).

Modulation of ADAR1's nuclear concentration:

Possible functions of the Modulator Of Import (MOI):

To allow nucleocytoplasmic shuttling of the longer form of ADAR1, hsADAR1i, this version of the protein contains a Crm1-dependent NES within the aminoterminal Z-DNA binding domain in addition to the NLS. However, although the export signal is missing from the short ADAR1c protein, it was demonstrated that also this version of the protein can be found in the cytoplasm. Seemingly, the first dsRBD of hsADAR1 assists the NLS to regulate the nuclear enzyme concentration. As dsRBD1 interferes with nuclear accumulation of a reporter construct containing dsRBD3 as an active NLS, it was termed Modulator of Import (MOI) (Strehblow et al., 2002).

The fact that cytoplasmic accumulation caused by dsRBD1 is both specific for the dsRBD3-NLS and requires active RNA-binding of dsRBD1 or 3 suggests that the MOI might act by masking the dsRBD3-resident NLS. Alternatively, dsRBD1 might be an anchor, which is bound by a cytoplasmic factor. Yet another possibility is that the MOI displays NES activity (Strehblow et al., 2002). Especially the latter alternative has become more interesting since it has been revealed that other dsRBD containing proteins, for instance ILF3 and Staufen2, escape from the nucleus with the help of Exp5 in an RNA- and RanGTP binding dependent manner (Gwizdek et al., 2004; Macchi et al., 2004). Therefore, it seemed likely that Exp5 influences the cellular localization of hsADAR1 and, thus, experiments have been performed to investigate this possible interaction.

Exp5 binds several dsRBDs in a RanGTP- and RNA-dependent manner:

Initially, to study direct interactions of dsRBDs with Exp5, it was tried to precipitate Exp5 out of cell lysates. Anti-hsADAR1 antibodies were immobilized on beads and incubated with total, nuclear or cytoplasmic HeLa lysate. In principle, the antibodies should retain endogenous hsADAR1 bound to Exp5 if an interaction had taken place. Bound Exp5 should then be detected by western blotting. However, no interaction between the two proteins of interest could be observed by this method. Nonspecific bands on the western blots prevented the detection of Exp5 and could not be abolished by changing the setup (data not shown). In addition, the levels of the tested proteins were very low and karyopherin-cargo interactions are comparatively short-lived making detection rather difficult. As a result, henceforth pull down assays were exclusively performed by using recombinantly expressed and purified proteins.

On the one hand, it was expected that a specific interaction of the export receptor and a fusion protein containing dsRBD1 and 3 would require RanGTP binding forming a trimeric complex. On the other hand, the interaction should also be dependent on RNA-binding, as the reporter construct assays had predicted.

In fact, IPs were performed using Exp5 and a fusion protein containing all three dsRBDs of hsADAR1 in the presence or absence of GTP-locked Ran and double-stranded RNA. To ensure complete removal of RNA that might have still been bound to the proteins after their purification, in some IPs they were treated with RNAse V1, specific for dsRNA, prior to the actual assay. Additionally, to test for specificity of an Exp5-MOI/dsRBD3 interaction, also pull down assays with each of the three dsRBDs were executed individually.

Strikingly, as illustrated in *Fig. 12 A*, Exp5 could efficiently bind dsRBD1-2-3 when dsRNA or/and RanGTP was present (*Fig.12A: dsRBD1-2-3+dsRNA+RanGTP; dsRBD1-2-3+dsRNA, dsRBD1-2-3+RanGTP*). In contrast, when the export factor was incubated only with the dsRBD containing protein (*Fig.12A: dsRBD1-2-3*), a weak band was generated on the western blot. Hence, the increase of signal intensity when RNA and/or RanGTP were present clearly proofs that an interaction of Exp5 and the fusion protein is both, RNA- and RanGTP binding dependent, essentially as shown for mammalian Staufen2 (Macchi et al., 2004). Consistently, this notion can be underscored by the fact that when both proteins had been previously treated with RNAse, no interaction took place at all *(Fig.12A: dsRBD1-2-3 (RNAse))*. Remarkably, however, the signal could be recovered when RanGTP was added *(Fig.12A: dsRBD1-2-3+RanGTP (RNAse))*.

Similar results were obtained by testing ILF3 dsRBD2 as a positive control. Although the affinity of Exp5 to the protein harboring three dsRBDs appeared generally stronger, binding to the positive control is also RanGTP- and RNA dependent. Despite the interaction seemed less influenced by the presence of RNA and can hardly be recognized on the blot, signals were stronger when RNA had been added to the IP *(Fig.12 B)*. Therefore, so far, the outcome of the experiment indicates a specific interaction of Exp5 and both dsRBD-containing proteins.

To find out which of the three dsRBDs of hsADAR1 was bound by the karyopherin in the first place, each individual dsRBD was tested in IPs, as described above. Strikingly, it turned out that the export receptor interacts with each dsRBD in the same RanGTP- and RNA-binding dependent pattern (*Fig.12 B and C*). *In vivo* dsRBD2 of ADAR1 had no effect on the intracellular localization of reporter constructs. However, in IPs with hsADAR1 dsRBD2 that was RNAse treated showed an interaction with Exp5 in the presence of RanGTP *(Fig.12 B: dsRBD2+RanGTP (RNAse))*. Additional immunoprecipitation assays revealed that Exp5 can also bind to the second dsRBD of hsADAR1 in the same RNA-binding dependent manner with almost the same strength as shown for the other dsRBDs (data not shown). Furthermore, also a fusion protein containing mutated versions of dsRBD1 and 3 of hsADAR1, unable to bind RNA, equally interacted with Exp5 in IPs (data not shown), even though the same construct accumulated in the nucleus *in vivo* in transfection based assays (Strehblow et al., 2002). On the other hand, this reporter protein behaved in import assays exactly as the unmutated version. It did not enter the nucleus when incubated with cytosol in contrast to the *in vivo* situation (data not shown). Yet, however, *in vitro* it has not formerly been proven that these recombinantly expressed mutated dsRBDs have no affinity to dsRNA at all. Therefore, it is possible that this protein still displays weak affinity to RNA under appropriate conditions.

In summary, pull down assays evidently show that presence of either dsRNA or RanGTP enhances the affinity to a dsRBD1-2-3 fusion protein and, thus, indicate a specific interaction between the dsRBDs and Exp5. However, the finding that Exp5 can bind each individual dsRBD tested with similar effectivity as well as the fact that the karyopherin can bind dsRNA by itself, also implies that Exp5 displays a common affinity to dsRBDs especially in the presence of RanGTP and RNA.

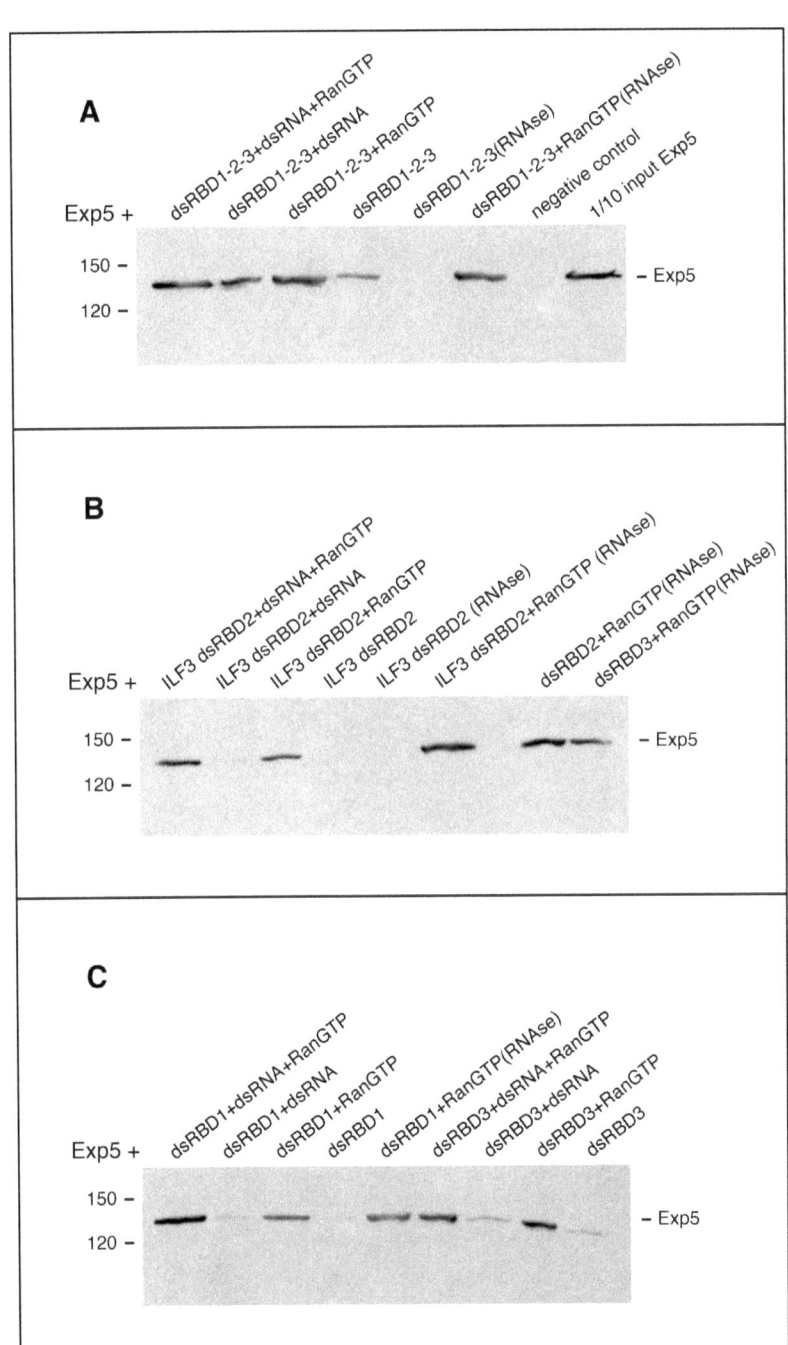

Figure 12: **RanGTP- and RNA-dependent interaction of Exp5 and several dsRBDs:**
(A) In pull down assays Exp5 can bind to a protein containing the three dsRBDs of hsADAR1. This interaction is seemingly RNA- and Ran-GTP binding dependent, as the intensities of the bands on the blot are higher in the presence of either of the two molecules (dsRBD1-2-3+dsRNA/+RanGTP). Interestingly, weak binding of the karyopherin to the fusion protein can also be observed when neither dsRNA nor RanGTP was added (dsRBD1-2-3). This interaction obviously results from residual RNA bound by the two proteins, which was proven by RNAse V1 treatment of the proteins previously to the IP (dsRBD1-2-3 (RNAse)). Notably, the interaction can be rescued in the presence of RanGTP (dsRBD1-2-3+RanGTP (RNAse)). As the negative control shows, Exp5 by itself does not generate any background when incubated with the beads (negative control) and in the last lane 10% Exp5 input was loaded onto the gel (1/10 input Exp5). (B) Although the band intensities in case of the positive control, dsRBD2 of hsILF3, are generally weaker than those with hsADAR1 dsRBD1-2-3, RNA- and RanGTP dependent binding of Exp5 to this dsRBD can also be noticed (ILF3 dsRBD2 +dsRNA/+RanGTP/(RNAse)). (B + C) To find out which of ADAR1's dsRBDs is preferentially bound by Exp5, each of the domains has been individually tested for interaction with the export receptor. As each dsRBDs displays an almost equal affinity to Exp5 in the same RanGTP- and RNA binding dependent manner, the outcome suggests a general interaction of the karyopherin and the three ADAR1 dsRBDs (dsRBD1/2/3+ dsRNA/+RanGTP/(RNAse)).

No indications for NES activity of the MOI *in vivo*:

Although almost every pyruvate kinase fusion reporter construct containing the first and the third dsRBD of hsADAR1 accumulates in the cytoplasm, even though a functional NLS is present, not every cell, however, overexpressing these reporter constructs localizes the fusion proteins exclusively to the cytoplasm. Interestingly, especially in cells expressing the artificial constructs in moderate levels, the proteins can also partially be found in the nucleus (*Fig.13: NLS38a, b*).

To analyze whether the expressing rate alters the localization of the fusion proteins indicating that the MOI effect derives from too high concentrations of the dsRBD-containing proteins leading to aggregation in the cytoplasm, a cell line stably expressing a construct containing all three dsRBDs of hsADAR1 was cultivated. But, no differences in the intracellular localization of the proteins could have been detected. In consistence with transfected cells overexpressing the construct, in some cases it was partially found in nucleus, but in the majority of the cells nuclear accumulation of the fusion proteins was interfered (*Fig.13: st.NLS38*). Therefore, no link between expression level and nucleocytoplasmic distribution of the constructs could be noticed underlining that cytoplasmic accumulation of reporter proteins is no artifact induced by overexpression. It rather suggests the biological importance of the MOI in hsADAR1 to regulate the nuclear concentration of the enzyme.

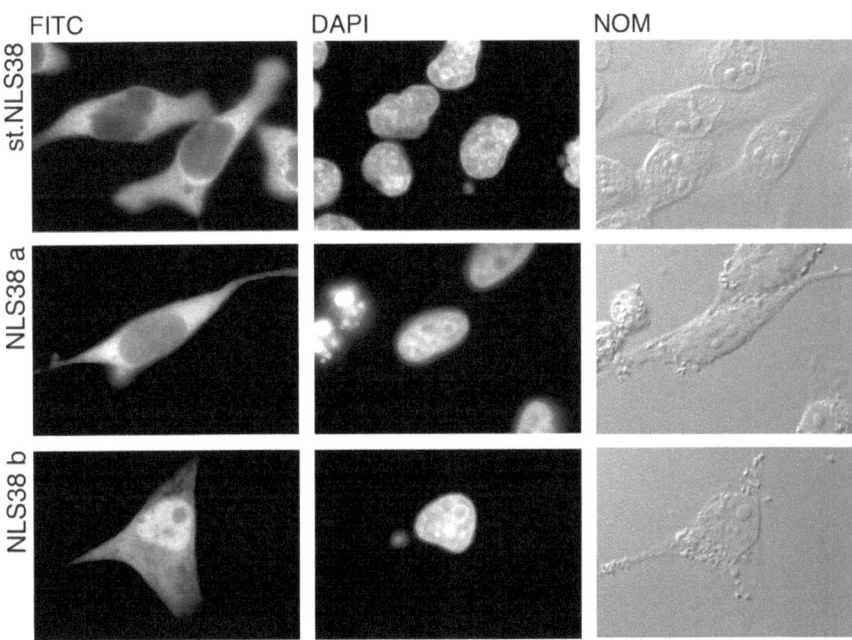

Figure 13: **Expression rate does not alter the localization of reporter proteins containing the MOI:**
Even though stably expressed in moderate levels, reporter proteins including the MOI and the NLS of hsADAR1 accumulate mostly in the cytoplasm (st.NLS38) and, as can be observed in cells overexpressing these proteins, also partially in the nucleus (NLS38a, b). This finding indicates that MOI induced interference with nuclear import is not an effect based on cytoplasmic aggregation of the protein due to artificially high protein amounts in the cell. It rather suggests that the first dsRBD of the enzyme indeed has a biological function in regulating the nuclear concentration of ADAR1.

Knock down of Exp5 by RNAi:

To study the effect on the nucleocytoplasmic distribution of hsADAR1 in human cells in the absence of Exp5, the karyopherin was knocked down by RNAi. By transfecting simultaneously two pairs of already published siRNAs the concentration of Exp5 in living HeLa cells was significantly reduced (Lund et al., 2003; Yi et al., 2004). In every experiment the level of the karyopherin, which is predominantly nuclear, was monitored by immunofluorescence using an Exp5 specific antibody (*Fig.14 A: endog. Exp5 vs. Exp5 RNAi*). In addition, the reduction of the export receptor was confirmed by western blots using total HeLa extracts of siRNA transfected and untransfected cells. To ensure that equal amounts of proteins had been loaded onto the gels, for control an anti-αActin antibody was used separately (*Fig.14 B*).

Nuclear transport is a saturable process. Theoretically, if Exp5 mediates nuclear export of hsADAR1 every reporter protein containing the first and third dsRBD should be able to enter the nucleus, but should be retained in this compartment in Exp5 depleted cells. Moreover, the distribution of endogenous hsADAR1, which can be found in the nucleus as well as in the cytoplasm, should be shifted towards the nucleus.

However, knock down of Exp5 caused surprisingly an even higher cytoplasmic concentration of endogenous hsADAR1 compared to wild type cells (*Fig.14 C: endog. ADAR1 vs. ADAR1 RNAi*). This result suggests that Exp5 acts as an import receptor rather than an export factor, as assumed. Additionally, no change in the localization of overexpressed or stably expressed reporter constructs containing dsRBD1-2-3 could be observed when the level of Exp5 was decreased, supporting the above statement (*Fig.14 C: NLS38 vs. NLS38 RNAi*).

As Exp5's yeast homologue Kap142p was shown to mediate transport in either direction across the nuclear pore, a possible function of Exp5 serving as the import receptor of ADAR1 was investigated. For that reason, Exp5 RNAi treated HeLa cells were transfected with a construct only containing dsRBD3. But, Exp5 knock down did not interfere with nuclear accumulation of this reporter protein (data not shown).

In total, Exp5 RNAi data did not prove a direct interaction of Exp5 and hsADAR1 *in vivo*. Nevertheless, due to the fact that the export factor was not knocked out completely, an involvement in the shuttling activity of hsADAR1 is still possible but does not seem likely.

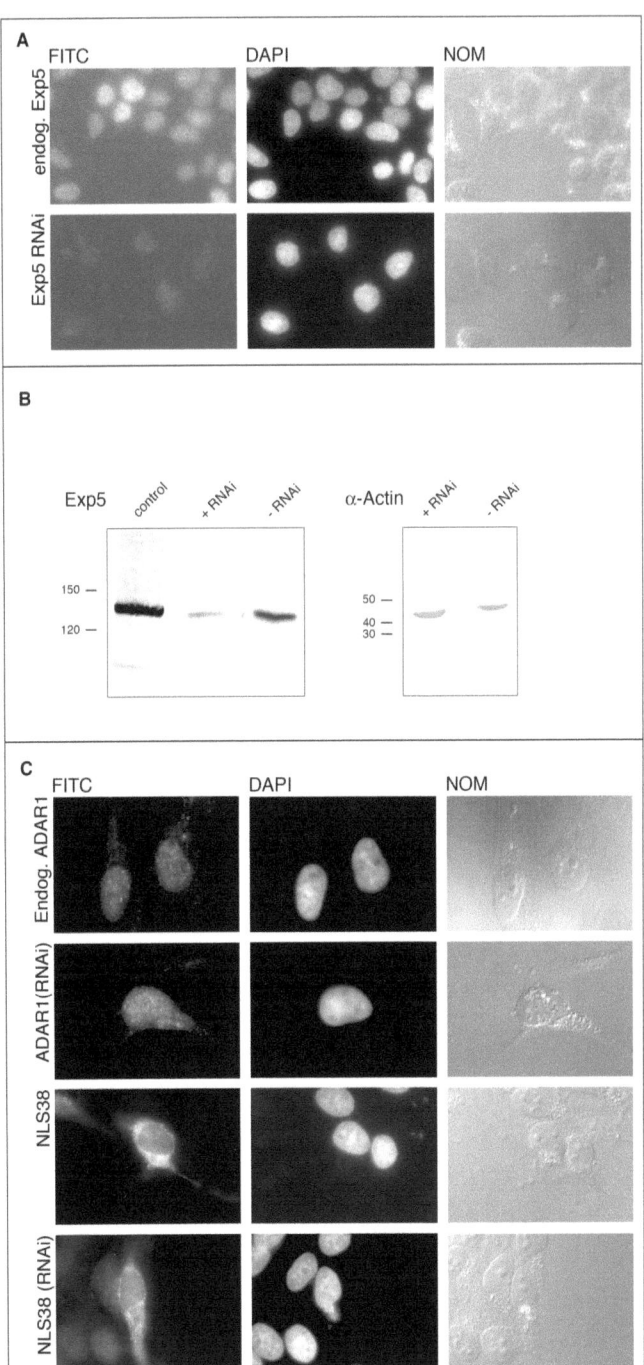

Figure 14: **Exp5 RNAi does not interfere with nuclear accumulation of hsADAR1:**
(A) Endogenous Exp5 is predominantly nuclear in HeLa cells (endog. Exp5). When transfected with Exp5 siRNAs, the expression rate of the export receptor can seemingly be reduced in tissue culture cells, but not erased completely (Exp5 RNAi). (B) Knock down of Exp5 was confirmed in total HeLa lysates by western blotting. To ensure that almost equal amounts of proteins were loaded onto the gel, an anti-α-Actin antibody was used to compare band intensities. (C) In contrast to endogenous hsADAR1, which is mostly nuclear (endog. ADAR1), in HeLas with less Exp5 concentrations the enzyme is almost randomly distributed throughout the cells giving rise that the karyopherin mediates rather import than export of hsADAR1 (ADAR1 RNAi). In consistence, a fusion protein containing the three dsRBDs of hsADAR1 does not move into the nucleus when Exp5 has been knocked down (NLS38, NLS38 RNAi). Bar = 20μm.

Ongoing experiments: heterokaryon assays, S35 labeled protein microinjection in the nuclei of *Xenopus* oocytes and affinity chromatography using dsRBD1-2-3 of hsADAR1:

As the experiments described above indicate that Exp5 does not facilitate transport of hsADAR1 across the nuclear pore, further experiments should clarify if the MOI is an NES interacting with a different export receptor or if it retains the enzyme in the cytoplasm to avoid harmful concentrations of ADAR1 in the nucleus. To address the question whether export or cytoplasmic tethering takes place, heterokaryon assays and microinjection experiments will be performed.

As previously mentioned, the short ADAR1 version, ADAR1c, can almost exclusively be found in the nucleus giving the impression that the protein is not able to escape from this compartment. Hypothetically, however, the nuclear signals obtained from transfection assays can also result from shuttling proteins that are imported much more efficiently than exported. Therefore, the ability of ADAR1c to leave the nucleus will be scrutinized by heterokaryon assays.

In principle, after stop of translation a tissue culture cell, expressing the reporter protein, will be fused with a wild type cell of a different species, for instance human and mouse, enabling discrimination of the two nuclei. The result is a large cell containing two nuclei in a joint cytoplasm. If ADAR1c shuttles, it will be exported from the original nucleus and will immediately be caught by an import receptor, which will interact with an NPC of either nucleus. Consequently, after some time the total amount of ADAR1c in the cell will be divided by the two nuclei. But, if the MOI does not display NES activity, ADAR1c will only be detected in the original nucleus.

Furthermore, with a similar intention microinjection experiments will be executed. In fact, a fusion protein containing all three dsRBDs of hsADAR1 will be injected into the comparably large nuclei of *Xenopus* oocytes. After an incubation period cytoplasm and nuclei will be separated and loaded onto protein gels. If the MOI can trigger export, the reporter protein will be detected in the cytoplasmic fraction on western blots.

As initial attempts failed, by which antibodies were used to detect the proteins, in future the assay will be performed using S35 radioactively labeled *in vitro* translated dsRBD-proteins to increase the detection limit.

Finally, efforts to find interaction partners of the MOI are currently undertaken. For that reason, affinity chromatography columns immobilizing the dsRBD1-2-3 fusion protein, similar as described above for dsRBD3 to identify the corresponding import receptor, are going to be prepared. In either case, nuclear export or cytoplasmic anchor, this method will rapidly help us to fish nuclear or cytosolic proteins that bind the three dsRBDs. By comparing the identified proteins with the dsRBD3-data, we will be able to exclude proteins that seemingly bind dsRBDs in common, leaving candidates exclusively interacting with the MOI-NLS containing protein.

Provided that karyopherins are found in the samples, they will be tested for their ability to mediate export of hsADAR1 by the methods portrayed above for Exp5. But, the interaction of any other cytoplasmic protein and dsRBD1-2-3 will first be confirmed by IPs and the candidate will ultimately knocked down by RNAi. Any intracellular localization change of a construct harboring the MOI and the NLS would point to a cytoplasmic anchoring function of the dsRBD binding protein of hsADAR1.

4. Discussion:

The development of intracellular membranes spatially separating two compartments in a cell was one of the most important steps in evolution. Generally, this invention enables that chemical gradients within a cell can be established, creating different environments, which allow the same sort of molecule to participate in different reactions at the same time. That includes that the membranes are also used to protect molecules from reacting with others and, thus, representing an alternative way to regulate the activity of proteins.

The complexity of the regulation of hsADAR1 obviously reflects its importance and diversity of functions in a cell, which spans from generating protein variants and modulating the miRNA silencing machinery to cellular defense against viral infections. As it has been shown for many other enzymes, nucleocytoplasmic shuttling is one level to control that ADAR1 edits its substrates at the right place and time. Seemingly, that is the reason why its RNA binding domains, which are essential for all of ADAR1's functions, harbor on the one hand the nuclear localization signal, but also a region that modulates import of the enzyme. By making the latter dependent on RNA-binding, the cell can apparently fine-tune the concentrations of the protein in either compartment ensuring that an optimal ratio between hsADAR1 and its substrates is maintained.

This study was set up to investigate the mechanisms underlying the regulation of nuclear transport of hsADAR. On the one hand, the atypical NLS overlapping the third dsRBD of the deaminase was extensively analyzed and a search for the karyopherin mediating import of the enzyme has been performed. On the other hand research was concentrated on the possible function of the MOI acting as an NES interacting with Exp5. Although, however, the major questions, objected in the introduction, could not have been solved completely, important approaches were realized.

Studying nuclear import of hsADAR1:

Identification of NLS-comprising residues within dsRBD3:

The search for amino acids comprising the dsRBD3-resident NLS turned out to be more difficult than assumed. Even though various methods were executed and finally 11 promising candidate amino acids were identified, the residues interacting with an import receptor could not be identified. Indeed, the candidates might be important for NLS activity, but they were not sufficient to trigger import when put on a different dsRBD, that normally does not accumulate in the nucleus.

As previously indicated, the key question has consequently to be asked whether not all NLS-comprising residues have been considered as candidates or whether the dsRBD folds into a unique 3D structure exposing the NLS-residues in a different way making them accessible to the import machinery. In the first case, it is possible that amino acids, which are not particularly conserved but are important for binding an RNA or are just covered when RNA is bound, can also interact with an import receptor. That seems to be a potential scenario considering the fact that it has been proven that dsRBD3 can enter the nucleus without RNA (Eckmann et al., 2001), but we do not know whether it can be imported when RNA is bound. Consistently, it would fit into the model of the MOI triggering NLS masking by recruiting RNAs. Hence, such a situation will be analyzed.

Whether the third dsRBD of hsADAR1 folds differently compared to other dsRBDs, we will know due to NMR studies in near future. Particularly, the C-terminal flanking region of the dsRBD will be focused on where two residues essential for NLS activity were identified and an almost perfect homology between dsRBD3 of the human and the amphibian ADAR1 protein exists. Furthermore, three online protein structure prediction programs predicted an additional 5-10 amino acids α-helix C-terminally of the dsRBD consensus sequence, as it was recently shown for the yeast homologue of RNAseIII, Rnt1p. In fact, NMR spectroscopy and X-ray crystallography revealed an extended dsRBD forming a third helix in the yeast endoribonuclease. The authors suggest that this helix allows the dsRBD of Rnt1p to bind short RNAs by changing the conformation of the N-terminal helix (Leulliot et al., 2004). In case of hsADAR1, an element modulating the structure of dsRBD3 could be the trigger enabling an import receptor to interact with the NLS.

Which import receptor mediates import of hsADAR1?

The second interesting part of this study is the search for an import receptor mediating nuclear import of hsADAR1. As reported above, several potential karyopherins have been tested in pull down and import assays for their ability to specifically bind and import the dsRBD-resident NLS. Interestingly, every single import factor generated at least weak signals in the nucleus of permeabilized HeLa cells. This unexpected phenomenon has already been described by Dirk Görlich and colleagues (Jäkel and Görlich, 1998; Jäkel et al., 2002). When added in abundant amounts in import assays several import receptors, such as TRN-1, Impα/β, Imp5 and Imp9, caused nuclear accumulation of the ribosomal protein S7. The authors argue that karyopherins serve a second purpose, namely by shielding basic cargoes against undesired interactions during transit to the nucleus. Many basic proteins tend to aggregate in an importin-free cytosol, apparently through multivalent ionic interactions with themselves or RNAs. That is the reason why it is believed that import receptors suppress aggregation of potential cargoes in polyanionic environment.

However, the outcome of the pull down and import assays indicate that none of the recombinantly expressed karyopherins mediates nuclear import of hsADAR1 *in vivo*. Nevertheless, affinity chromatography experiments using dsRBD3 as bait revealed Transportin-1 as the most promising candidate. In addition, this transport receptor was shown to mediate import of several proteins involved in RNA metabolism via binding to atypical NLS-sequences that do not essentially share consensus, as reported in the introduction. For instance, hnRNP A1 contains a 38 amino acid M9 domain, whereas transport of RNA helicase A is mediated by a 110 amino acid sequence in the C-terminus (reviewed in Michael, 2000).

As initial tests give the impression that this receptor triggers import of dsRBD3 at least 20-fold more efficient than the other importins together with the fact that the karyopherin had been found to interact with several atypical NLSs folding into complex 3D structures, we are currently focusing on TRN-1. Should the experiments mentioned above confirm our preliminary data, the import receptor will be ultimately knocked down by RNAi to see if endogenous hsADAR1 and transiently transfected reporter constructs are still able to get into the nucleus. Pursuing this strategy the question whether TRN-1 is the corresponding import factor of hsADAR1's NLS, will be clarified soon.

What is the function of the MOI?

The first dsRBD of hsADAR1 is a cis-acting element regulating the nuclear concentration of the enzyme in an RNA-binding dependent manner. In fact, the MOI interferes with nuclear accumulation when present in reporter proteins containing the ADAR's NLS. For that reason two possibilities to oppose nuclear import have been considered. The MOI either inhibits import or stimulates export.

Interestingly, both scenarios have been described for the mammalian Staufen protein family that is involved in RNA localization in polarized neurons. As reported above, Stau2 is a predominantly cytoplasmic protein that can enter the nucleus via a bipartite NLS in the spacing region between the third and fourth dsRBD. Remarkably, Exp5 was shown to interact with Stau2 when RNA is bound to dsRBD3 *in vitro*, but not with a mutated dsRBD defective in RNA binding. In addition, in Exp5 downregulated cells only the largest of the three known isoforms of Stau2 accumulates in the nucleus, strongly indicating that nuclear exit of this Stau2 isoform is mediated by Exp5 (Macchi et al., 2004).

In contrast, the nucleocytoplasmic shuttling activity of Stau1 seems to be regulated differently. Indeed, as shown for Stau2, a dsRBD of Stau1 displays MOI activity that appears to be based on RNA-binding of the dsRBD supported by a second dsRBD. But although import of the protein is also triggered by a bipartite NLS C-terminally of the third dsRBD, which is most important for RNA binding of the protein, and even though Exp5 can interact with this dsRBD in an RNA-binding dependent manner (Brownawell and Macara, 2002), knock down of Exp5 did not result in nuclear accumulation of Stau1. Therefore, it is rather suggested that transport into the nucleus is regulated by mechanisms that involve cytoplasmic retention of the protein (Martel et al., 2005).

Similarly, several experiments described herein argue that cytoplasmic accumulation of hsADAR1 reporter constructs containing the MOI is caused by interfering with nuclear import rather than export mediated by Exp5. Although, the export receptor interacts with the enzyme in the presence of dsRNA and RanGTP in pull down assays, the fact that it binds every single dsRBD tested with almost equal strength indicates a general affinity of Exp5 to dsRBDs.

However, dsRNA and/or RanGTP enhances the interaction of the export receptor with the dsRBDs. Interestingly, especially in case of dsRBD1-3, to generate an intense band either RNA or RanGTP have to be present. But, when added both in same reaction, the signal was not increased. Hence, if Exp5 is indeed the corresponding export factor transporting ADAR1 out of the nucleus, it would be one possibility that the outcome of the IPs indicates that export is mediated by two competing mechanisms. Both are dependent on Exp5- and RanGTP binding, but in the first scenario, dsRNA-substrates are specifically bound and transported out of the nucleus. On the other hand, the pull down assays with RNAse treated proteins showed a strong interaction between Exp5 and dsRBD1-3 in the presence of only RanGTP. This fact suggests that the export receptor could also export the enzyme without RNA binding. Interestingly, this model matches previous observations regarding Exp5's binding requirements. When the karyopherin was discovered, the authors claimed that Exp5 interaction with dsRBDs is prevented when RNA is bound (Brownawell and Macara, 2002). However, shortly after it was proposed that the export receptor can indeed also bind dsRBDs when associated with RNA. Furthermore, Exp5 can bind minihelix-containing RNA molecules by itself (Gwizdek et al., 2003). Therefore, further experiments will shed more light on this model.

Nevertheless, Exp5 RNAi did not lead to nuclear accumulation of endogenous hsADAR1 and dsRBD1-3 containing reporter proteins. On the contrary, Exp5 siRNA transfection even caused a higher cytoplasmic concentration of the enzyme, giving rise to an import function of the karyopherin. However, evidences for Exp5 mediating import of hsADAR1 have not been obtained.

But, MOI acting as an NES cannot be excluded, even though Exp5 does not seem likely to trigger export of the enzyme. As mentioned before, heterokaryon and microinjection assays that are currently performed will clarify whether the short ADAR1 version can potentially exit the nucleus.

If it turns out that the MOI does not display NES activity, investigations will be concentrated on the question whether cytoplasmic retention is achieved through protein-protein interaction or NLS masking via RNA-binding of the first and the third dsRBD of ADAR.

The first case, a cytoplasmic anchor mechanism, has already been described for the transcription factor NFκB. All members of the NFκB family contain an approximately 300 amino acid sequence, which is important for DNA-binding, dimerization and interaction with the inhibitor protein IκB. If the transcription factor is not needed in the nucleus, it accumulates in the cytoplasm, where it is bound as a homo- or heterodimer by the inhibitor. Hence, access to the NLS is blocked and NFκB cannot enter the nucleus (Beg et al., 1992; Ganchi et al., 1992; Latimer et al., 1998).

In fact, interactions between ADAR1 and other proteins in the cytoplasm do not seem unlikely, since several dsRBDs have been implicated in multimerization, as reported in the introduction (reviewed in Saunders and Barber, 2003). But also, homodimerization of hsADAR1 cannot be excluded as the mechanism causing interference with nuclear accumulation of fusion constructs containing the first and the third dsRBD of the protein. Consequently, to find potential interaction partners of the MOI, affinity chromatography columns with the three dsRBDs of the human enzyme and subsequent mass spectrometry analysis will be carried out.

On the other hand, the MOI effect could also be based on competition between interaction of an import receptor and the dsRBD3-resident NLS, and simultaneous binding of a common RNA by dsRBD1 and 3 in the cytoplasm, essentially as shown for the human RNA helicase A. Although this enzyme contains a nuclear transport domain in the C-terminus, an interplay between both N-terminal dsRBDs of RHA seems to tether the protein in the cytoplasm (Aratani et al., 2001; Fujita et al., 2005).

Theoretically, regulating ADAR1 via NLS masking is compatible with every observation mentioned above regarding the shuttling activity of the enzyme and, in addition, it would clarify some open questions. First, NLS masking goes along with the fact that the intracellular localization of endogenous hsADAR1 is shifted towards the cytoplasm when the cells are transfected with Exp5 siRNAs. It is possible that more ADAR proteins are excluded from the nucleus due to the presence of additional potential substrates rather than the absence of the export receptor. However, control experiments have not confirmed this situation, yet.

Furthermore, if the NLS of ADAR1 is only active when there is no RNA bound, it would explain why we could not identify NLS-comprising residues. Therefore, efforts will be undertaken to clarify the question whether NLS masking is the mechanism underlying the MOI effect.

5. References:

Aaronson, R.P., and Blobel, G. 1974. On the attachment of the nuclear pore complex. *J Cell Biol.* Vol.62(3):746-754

Anant, S., and Davidson, N.O. 2001. Molecular mechanisms of apolipoprotein B mRNA editing. *Curr Opin Lipidol.* Vol.12(2):159-165

Aratani, S., Fujii, R., Oishi, T., Fujita, H., Amano, T., Ohshima, T., Hagiwara, M., Fukamizu, A., and Nakajima, T. 2001. Dual roles of RNA helicase A in CREB-dependent transcription. *Mol Cell Biol.* Vol.21:4460-4469

Athanasiadis, A., Rich, A., and Maas, S. 2004. Widespread A-to-I RNA editing of Alu-containing mRNAs in the human transcriptome. *PLoS Biol.* Vol.2(12):e391

Bachi, A., Braun, I.C., Rodrigues, J.P., Pante, N., Ribbeck, K., von Kobbe, C., Kutay, U., Wilm, M., Görlich, D., Carmo-Fonseca, M., and Izaurralde, E. 2000. The C-terminal domain of TAP interacts with the nuclear pore complex and promotes export of specific CTE-bearing RNA substrates. *RNA.* Vol.6:136-158

Bartel, D.P. 2004. MicroRNAs: genomics, biogenesis, mechanism, and function. *Cell.* Vol.116:281-297

Barth, C., Greferath, U., Kotsifas, M. and Fisher, P.R. 1999. Polycistronic transcription and editing of the mitochondrial small subunit (SSU) ribosomal RNA in Dictyostelium discoideum. *Curr Genet.* Vol.36 (1-2): 55-61

Bass, B.L., and Weintraub, H.A. 1987. Developmental regulated activity that unwinds RNA duplexes. *Cell.* Vol.48:607-613

Bass, B.L., and Weintraub, H.A. 1988. An unwinding activity that covalently modifies its dsRNA substrate. *Cell.* Vol.55:1089-1098

Bass, B.L., Nishikura, K., Keller, W., Seeburg, P.H., Emeson, R.B., O'Connell, M.A., Samuel, C.E., and Herbert, A. 1997. A standardized nomenclature for adenosine deaminases that act on RNA. *RNA.* Vol.3:947-949

Bass, B.L. 1997. RNA editing and hypermutation by adenosine deamination. *TIBS.* Vol.22:157-162

Bass, B.L. 2002. RNA editing by adenosine deaminases that act on RNA. *Annu Rev Biochem.* Vol.71:817-846

Bayliss, R., Littlewood, T., and Steward, M. 2000. Structural basis for the interaction between FxFG nucleoporin repeats and importin-beta in nuclear trafficking. *Cell.* Vol.102:99-108

Bear, J., Tan, W., Zolotukhin, A.S., Tabernero, C., Hudson, E.A., and Felber, B.K. 1999. Identification of Novel Import and Export Signals of Human TAP, the Protein That Binds to the Constitutive Transport Element of the Type D Retrovirus mRNAs. *Mol Cell Biol.* Vol.19:6306-6317

Beg, A.A., Ruben, S.M., Scheinman, R.I., Haskill, S., Rosen, C.A., and Baldwin, A.S. Jr. 1992. I Kappa B interacts with the nuclear localisation sequence of the subunits of NF Kappa B: a mechanism for cytoplasmic retention. *Genes Dev.* Vol.12B:2664-2665

Benne, R., Van den Burg, J., Brakenhoff, J.P., Sloof, P., Van Boom, J.H., and Tromp, M.C. 1986. Major transcript of the frameshifted coxII gene from trypanosome mitochondria contains four nucleotides that are not encoded in the DNA. *Cell.* Vol.46 (6):819-826

Ben-Shlomo, H., Levitan, A., Shay, N.E., Goncharov, I., and Michaeli, S. 1999. RNA editing associated with the generation of two distinct conformations of the trypanosomatid Leptomonas collosoma 7SL RNA. *J Biol Chem.* Vol.274(36):25642-25650

Bischoff, F.R., and Ponstingl, H. 1991. Catalysis of guanine nucleotide exchange on Ran by the mitotic regulator RCC1. *Nature.* Vol.354:80-82

Bischoff, F.R., Klebe, C., Kretschmer, J., Wittinghofer, A., and Ponstingl, H. 1994. RanGAP1 induces GTPase activity of nuclear Ras-related Ran. *Proc Natl Acad Sci. USA.* Vol.91:2587-2591

Bischoff, F.R., Krebber, H., Smirnova, E., Dong, W., and Ponstingl, H. 1995. Co-activation of RanGTPase and inhibition of GTP dissociation by Ran-GTP binding protein RanBP1. *EMBO J.* Vol.14:705-715

Blanc, V., and Davidson, N.O. 2003. C-to-U RNA editing: mechanisms leading to genetic diversity. *J Biol Chem.* Vol.278(3):1395-1398

Bohnsack, M.T., Regener, K., Schwappach, B., Saffrich, R., Paraskeva, E., Hartmann, E., and Görlich, D. 2002. Exp5 exports eEF1A via tRNA from nuclei and synergizes with other transport pathways to confine translation to the cytoplasm. *EMBO J.* Vol. 1(22):6205-6215

Bohnsack, M.T., Czaplinski, K., and Görlich, D. 2004. Exportin 5 is a RanGTP-dependent dsRNA-binding protein that mediates nuclear export of pre-miRNAs. *RNA.* Vol.10(2):185-191

Bonifaci, N., Moroianu, J., Radu, A., and Blobel, G. 1997. Karyopherin beta2 mediates nuclear import of a mRNA binding protein. *Proc Natl Acad Sci. USA,* Vol.94:5055-5060

Brooks, R., Eckmann, C.R., and Jantsch, M.F. 1998. The double-stranded RNA-binding domains of *Xenopus laevis* ADAR1 exhibit different RNA-binding behaviors. *FEBS Lett.* Vol.434:121-126

Brown, B.A., Lowenhaupt, K., Wilbert, C.M., Hanlon, E.B., and Rich, A. 2000. The Z-alpha domain of the editing enzyme dsRNA adenosine deaminase binds left-handed Z-RNA as well as Z-DNA. *Proc Natl Acad Sci USA.* Vol.97(25):13532-13536

Brownawell, A., and Macara, I.G. 2002. Exportin 5, a novel karyopherin, mediates nuclear export of double-stranded RNA binding proteins. *J Cell Biol.* Vol.156(1):53-64

Brusa, R., Zimmermann, F., Koh, D.S., Feldmeyer, D., Gass, P., Seeburg, P.H., and Sprengel, R. 1995. Early-onset epilepsy and postnatal lethality associated with an editing-deficient GluR-B allele in mice. *Science.* Vol.270(5242):1677-1680

Brzostowski, J., Robinson, C., Orford, R., Elgar, S., Scarlett, G., Peterkin, T., Malartre, M., Kneale, G., Wormington, M., and Guille, M. 2000. RNA-dependent cytoplasmic anchoring of a transcription factor subunit during Xenopus development. *EMBO J.* Vol.19(14):3683-3693

Burns, C.M., Chu, H., Rueter, S.M., Hutchinson, L.K., Canton, H., Sanders-Bush, E., and Emeson, R.B. 1997. Regulation of serotonin-2C-receptor G-protein coupling by RNA editing. *Nature*. Vol.387:303-308

Bycroft, M., Grunert, S., Murzin, A.G., Proctor, M., and St Johnston, D. 1995. NMR solution structure of a dsRNA binding domain from Drosophila staufen protein reveals homology to the N-terminal domain of ribosomal protein S5. *EMBO J*. Vol.14(14):3563-3571. Erratum in: *EMBO J*. Vol.14(17):4385

Calado, A., Treichel, N., Müller, E.C., Otto, A., and Kutay, U. 2002. Exportin-5-mediated nuclear export of eukaryotic elongation factor 1A and tRNA. *EMBO J*. Vol.21(22):6216-6224

Callan, H.G., and Tomlin, S.G. 1950. Experimental studies on amphibian oocyte nuclei. I. Investigation of the structure of the nuclear membrane by means of the electron microscope. *Proc R Soc Lond B Biol Sci*. Vol.137(888):367-378

Carlson, C.B., Stephens, O.M., and Beal, P.A. 2003. Recognition of double-stranded RNA by proteins and small molecules. *Biopolymers*. Vol.70(1):86-102

Cattaneo, R., Kaelin, K., Baczko, K., and Billeter, M.A. 1989. Measles virus editing provides an additional cysteine-rich protein. *Cell*. Vol.56(5):759-764

Chan, L. 1992. Apolipoprotein B, the major protein component of triglyceride-rich and low density lipoproteins. *J Biol Chem*. Vol.267(36):25621-25624

Chang, F.L., Chen, P.J., Tu, S.J., Wang, C.J., and Chen, D.S. 1991. The large form of hepatitis delta antigen is crucial for assembly of hepatitis delta virus. *Proc Natl Acad Sci. USA*. Vol.88:8490-8494

Chen, S.H., Habib, G., Yang, C.Y., Gu, Z.W., Lee, B.R., Weng, S.A., Silberman, S.R., Cai, S.J., Deslypere, J.P., and Rosseneu, M. 1987. Apolipoprotein B-48 is the product of a messenger RNA with an organ-specific in-frame stop codon. *Science*. Vol.238(4825):363-366

Chen, C.X., Cho, D.S., Wang, Q., Lai, F., Carter, K.C., and Nishikura, K. 2000. A third member of the RNA-specific adenosine deaminase gene family, ADAR3, contains both single- and double-stranded RNA binding domains. *RNA*. Vol.6(5):755-767

Chester, A., Scott, J., Anant, S., and Navaratnam, N. 2000. RNA editing: cytidine to uridine conversion in apolipoprotein B mRNA. *Biochim Biophys Acta.* Vol.1494(1-2):1-13

Chi, N.C., and Adam, S.A. 1997. Functional domains in nuclear import factor p97 for binding the nuclear localization sequence receptor and the nuclear pore. *Mol Biol Cell.* Vol.8:945-956

Cho, D.S., Yang, W., Lee, J.T., Shiekhattar, R., Murray, J.M., and Nishikura, K. 2003. Requirement of dimerization for RNA editing activity of adenosine deaminases acting on RNA. *J Biol Chem.* Vol.278(19):17093-17102

Conti, E., Uy, M., Leighton, L., Blobel, G., and Kuryan, J. 1998. Crystallographic Analysis of the Recognition of a Nuclear Localization Signal by the Nuclear Import Factor Karyopherin alpha. *Cell.* Vol.94:193-204

Cosentino, G.P., Venkatesan, S., Serluca, F.C., Green, S.R., Mathews, M.B., and Sonenberg, N. 1995. Double-stranded-RNA-dependent protein kinase and TAR RNA-binding protein form homo- and heterodimers in vivo. *Proc Natl Acad Sci USA.* Vol.92(21):9445-9449

Cronshaw, J.M., Krutchinsky, A.N., Zhang, W., Chait, B.T., and Matunis, M.J. 2002. Proteomic analysis of the mammalian nuclear pore complex. *J Cell Biol.* Vol.158: 915-927

Cullen, B.R. 2003. Nuclear mRNA export: insights from virology. *Trends Biochem Sci.* Vol.28(8):419-424

Dabiri, G.A, Lai, F., Drakas, R.A., and Nishikura, K. 1996. Editing of the GLuR-B ion channel RNA in vitro by recombinant double-stranded RNA adenosine deaminase. *EMBO J.* Vo.15(1):34-45

Desterro, J.M., Keegan, L.P., Lafarga, M., Berciano, M.T., O'Connell, M., and Carmo-Fonseca, M. 2003. Dynamic association of RNA-editing enzymes with the nucleolus. *J Cell Sci.* Vol.116:1805-1818

Doyle, M., and Jantsch, M.F. 2002. New and old roles of the double-stranded RNA-binding domain. *J Struct Biol.* Vol.140(1-3):147-153

Doyle, M., and Jantsch, M.F. 2003. Distinct in vivo roles for double-stranded RNA-binding domains of the Xenopus RNA-editing enzyme ADAR1 in chromosomal targeting. *J Cell Biol.* Vol.161(2):309-319

Dworetzky, S.I., Lanford, R.E., and Feldherr, C.M. 1988. The effects of variations in the number and sequence of targeting signals on nuclear uptake. *J Cell Biol.* Vol.107:1279–1287

Eckmann, C.R., and Jantsch, M.F. 1997. Xlrbpa, a double-stranded RNA-binding protein associated with ribosomes and heterogeneous nuclear RNPs. *J Cell Biol.* Vol.138(2):239-253

Eckmann, C.R., and Jantsch, M.F. 1999. The RNA-editing enzyme ADAR1 is localized to the nascent ribonucleoprotein matrix on Xenopus lampbrush chromosomes but specifically associates with an atypical loop. *J Cell Biol.* Vol.144(4):603-615

Eckmann, C.R., Neunteufl, A., Pfaffstetter, L., and Jantsch, M.F. 2001. The human but not the Xenopus RNA-editing enzyme ADAR1 has an atypical nuclear localisation signal and displays the characteristics of a shuttling protein. *Mol Biol Cell.* Vol.12:1911-1924

Estevez, A.M., and Simpson, L. 1999. Uridine insertion/deletion RNA editing in trypanosome mitochondria-a review. *Gene.* Vol.240(2):247-260

Fahrenkrog, B., and Aebi, U. 2003. The nuclear pore complex: nucleocytoplasmic transport and beyond. *Nat Rev Mol Cell Biol.* Vol.4:757-766

Fan, X.C., and Steitz, J.A. 1998. HNS, a nuclear-cytoplasmic shuttling sequence in HuR. *Proc Natl Acad Sci USA.* Vol.95:15293-15298

Fanara, P., Hodel, M.R., Corbett, A.H., and Hodel, A.E. 2000. Quantitative analysis of nuclear localization signal (NLS)-importin alpha interaction through fluorescence depolarization. Evidence for auto-inhibitory regulation of NLS binding. *J Biol Chem.* Vol.275:21218-21223

Fornerod, M., Ohno, M., Yoshida, M., and Mattaj, I.W. 1997. CRM1 is an export receptor for leucine-rich nuclear export signals. *Cell.* Vol.90(6):1051-1060

Fujita, H., Ohshima, T., Oishi, T., Aratani, S., Fujii, R., Fukamizu, A., and Nakajima, T. 2005. Relevance of nuclear localization and functions of RNA helicase A. *Int J Mol Med.* Vol.15:555-560

Gallo, A., Keegan, L.P., Ring, G.M., and O'Connell, M.A. 2003. An ADAR that edits transcripts encoding ion channel subunits functions as a dimer. *EMBO J.* Vol.22(13):3421-3430

Gallouzi, I.E., and Steitz J.A. 2001. Delineation of mRNA export pathways by the use of cell-permeable peptides. *Science.* Vol.294(5548):1895-1901. Erratum in: *Science.* Vol.296(5565):47

Ganchi, P.A., Sun, S.C., Greene, W.C., and Ballard, D.W. 1992. IκB/MAD-3 masks the nuclear localisation signal of NFκB p65 and requires the transactivation domain to inhibit NFκB p65 DNA binding. *Mol Biol Cell.* Vol.3:1339-1352

George, C.X., and Samuel, C.E. 1999. Human RNA-specific adenosine deaminase ADAR1 transcripts possess alternative exon 1 structures that initiate from different promoters, one constitutively active and the other interferon inducible. *Proc Natl Acad Sci. USA* Vol. 96:4621-4626

George, C.X., Wagner, M.V., and Samuel, C.E. 2005. Expression of interferon-inducible RNA adenosine deaminase ADAR1 during pathogen infection and mouse embryo development involves tissue-selective promoter utilization and alternative splicing. *J Biol Chem.* Vol.280(15):15020-15028

Gerber, A., Grosjean, H., Melcher, T., and Keller, W. 1998. Tad1p, a yeast tRNA-specific adenosine deaminase, is related to the mammalian pre-mRNA editing enzymes ADAR1 and ADAR2. *EMBO J.* Vol.1:4780-4789

Gerber, A.P., and Keller, W. 1999. An adenosine deaminase that generates inosine at the wobble position of tRNAs. *Science.* Vol.286:1146-1149

Gerber, A.P., and Keller, W. 2001. RNA editing by base deamination: more enzymes, more targets, new mysteries. *TIBS.* Vol.26(6):376-384

Görlich, D., Prehn, S., Laskey, R.A., and Hartmann, E. 1994. Isolation of a protein that is essential for the first step of nuclear protein import. *Cell.* Vol.79:767-778

Görlich, D., Kostka, S., Kraft, R., Dingwall, C., Laskey, R.A., Hartmann, E., and Prehn, S. 1995. Two different subunits of importin cooperate to recognize nuclear localization signals and bind them to the nuclear envelope. *Curr Biol.* Vol.5:383–392.

Görlich, D., Pante, N., Kutay, U., Aebi, U., and Bischoff, V.R. 1996. Identification of different roles for RanGDP and RanGTP in nuclear protein import. *EMBO J.* Vol.15:5584-5594

Görlich, D. 1998. Transport into and out of the cell nucleus. *EMBO J.* Vol.17:2721–2727

Görlich, D., Seewald, M.J., and Ribbeck, K., 2003. Characterization of Ran-driven cargo transport and the RanGTPase system by kinetic measurements and computer simulation. *EMBO J.* Vol.22:1088-1100

Gott, J.M., and Emeson, R.B. 2000. Functions and mechanisms of RNA editing. *Annu Rev Genet.* Vol.34:499-531

Gruss, O.J., Carazo-Salas, R.E., Schatz, C.A., Guarguaglini, G., Kast, J., Wilm, M., Le Bot, N., Vernos, I., Karsenti, E., and Mattaj, I.W. 2001. Ran induces spindle assembly by reversing the inhibitory effect of importin alpha on TPX2 activity. *Cell.* Vol.104:83-93

Gruter, P., Tabernero, C., von Kobbe, C., Schmitt, C., Saavedra, C., Bachi, A., Wilm, M., Felber, B.K., and Izaurralde, E. 1998. TAP, the human homolog of Mex67p, mediates CTE-dependent RNA export from the nucleus. *Mol Cell.* Vol.1(5):649-659

Gurevich, I., Tamir, H., Arango, V., Dwork, A.J., Mann, J.J., and Schmauss, C. 2002. Altered editing of serotonin 2C receptor pre-mRNA in the prefrontal cortex of depressed suicide victims. *Neuron.* Vol.34:349-356

Gwizdek, C., Ossareh-Nazari, B., Brownawell, A.M., Doglio, A., Bertrand, E., Macara, I.G., and Dargemont, C. 2003. Exportin-5 mediates nuclear export of minihelix-containing RNAs. *J Biol Chem.* Vol.278(8):5505-5508

Gwizdek, C., Ossareh-Nazari, B., Brownawell, A., Evers, S., Macara, I.G., and Dargemont, C. 2004. Minihelix-containing RNAs mediate exportin 5 nuclear export of the double-stranded RNA binding protein ILF3. *J Biol Chem.* Vol.279(2):884-91

Hajjar, A.M., and Linial, M.L. 1995. Modification of retroviral RNA by double-stranded RNA adenosine deaminase. *J Virol.* Vol. 69(9):5878-5882

Hartner, J.C., Schmittwolf, C., Kispert, A., Müller, A.M., Higuchi, M., and Seburg, P.H. 2003. Liver Disintegration in the Mouse Embryo Caused by Deficiency in the RNA-editing Enzyme ADAR1. *J Biol Chem.* Vol.279:4894-4902

Herbert, A., Alfken, J., Kim, Y.G., Mian, I.S., Nishikura, K., and Rich, A. 1997. A Z-DNA binding domain present in the human editing enzyme, double-stranded RNA adenosine deaminase. *Proc Natl Acad Sci USA.* Vol.94(16):8421-8426

Herbert, A., and Rich, A. 2001. The role of binding domains for dsRNA and Z-DNA in the *in vivo* editing of minimal substrates by ADAR1. *Proc Natl Acad Sci USA*, Vol.98(21):12132-12137

Herold, A., Truant, R., Wiegand, H., and Cullen, B.R.1998. Determination of the functional domain organization of the importin alpha nuclear import factor. *J Cell Biol.* Vol.143:309-318

Hetzer, M., Bilbao-Cortes, D., Walther, T.C., Gruss, O.J., and Mattaj, I.W. 2000. GTP hydrolysis by Ran is required for nuclear envelope assembly. *Mol Cell.* Vol.5:1013-1024

Higuchi, M., Maas, S., Single, F.N., Hartner, J., Rozov, A., Burnashev, N., Feldmeyer, D., Sprengel, R., and Seeburg, P.H. 2000. Point mutation in an AMPA receptor gene rescues lethality in mice deficient in the RNA-editing enzyme ADAR2. *Nature.* Vol.6:78-81

Hitti, E.G., Sallacz, N.B., Schoft, V.K., and Jantsch, M.F. 2004. Oligomerization activity of a double-stranded RNA-binding domain. *FEBS Lett.* Vol.574(1-3):25-30

Holley, R.W., Everett, G.A., Madison, J.T., and Zamir, A. 1965 Nucleotide sequence in yeast alanine transfer RNA. *J Biol Chem.* Vol.240:2122-2127

Hough RF, Bass BL. 1994. Purification of the Xenopus laevis double-stranded RNA adenosine deaminase. *J Biol Chem.* Vol.269(13):9933-9939

Hough, R.F., and Bass, B.L. 1997. Analysis of Xenopus dsRNA adenosine deaminase cDNAs reveals similarities to DNA methyltransferases. *RNA*. Vol.3:356-370

Hussain, M.M., Kancha, R.K., Zhou, Z., Luchoomun, J., Zu, H., and Bakillah, A. 1996. Chylomicron assembly and catabolism: role of apolipoproteins and receptors. *Biochim Biophsy Acta*. Vol.1300 (3):151-170

Hutvagner, G., and Zamore, P.D. 2002. A microRNA in a multiple-turnover RNAi enzyme complex. *Science*. Vol.293:834-838

Izaurralde, E., Jarmolowski, A., Beisel, C., Mattaj, I.W. Dreyfuss, G., and Fischer, U. 1997. A role for the M9 transport signal of hnRNPA1 in mRNA nuclear export. *J Cell Biol*. Vol.137 Nr.1:27-35

Jäkel, S., and Görlich, D. 1998. Importin β, transportin, RanBP5 and RanBP7 mediate nuclear import of ribosomal proteins in mammalian cells. *EMBO J*. Vol.17:4491-4502

Jäkel, S., Albig, W., Kutay, U., Bischoff, F.R., Schwamborn, K., Doenecke, D., and Görlich, D. 1999. The importin beta/importin 7 heterodimer is a functional nuclear import receptor for histone H1. *EMBO J*. Vol.18:2411–2423

Jäkel, S., Mingot, J.-M., Schwarzmaier, P., Hartmann, E., and Görlich, D. 2002. Importins fulfil a dual function as nuclear import receptors and cytoplasmic chaperons for exposed basic domains. *EMBO J*. Vol.21:377-386

Kalab, P., Pu, R.T., and Dasso, M. 1999. The Ran GTPase regulates mitotic spindle assembly. *Curr biol*. Vol.9:481-484

Kalderon, D., Richardson, W.D., Markham, A.F., and Smith, A.E. 1984. Sequence requirements for nuclear location of simian virus 40 large-T antigen. *Nature*. Vol.311(5981):33-38

Källman, A.M., Sahlin, M., and Öhman, M. 2003. ADAR2 A-->I editing: site selectivity and editing efficiency are separate events. *Nucleic Acids Res*. Vol.31(16):4874-4881

Kang, Y., and Cullen, B.R. 1999. The human Tap protein is a nuclear mRNA export factor that contains novel RNA-binding and nucleocytoplasmic transport sequences. *Genes Dev.* Vol.13:1126-1139

Keegan, L.P., Gerber, A.P., Brindle, J., Leemans, R., Gallo, A., Keller, W., and O'Connell, M.A. 2000. The properties of a tRNA-specific adenosine deaminase from *Drosophila melanogaster* support an evolutionary link between pre-mRNA editing and tRNA modification. *Mol And Cell Biol.* Vol.20:825-833

Keegan, L.P., Gallo, A., and O'Connell, M.A. 2001. The many roles of an RNA editor. *Nature Rev.* Vol.2:869-878

Kehlenbach, R.H., Dickmanns, A., Kehlenbach, A., Guan, T., and Gerace, L. 1999. A role for RanBP1 in the release of CRM1 from the nuclear pore complex in a terminal step of nuclear export. *J. Cell Biol.* Vol.145:645–657

Kharrat, A., Macias, M.J., Gibson, T.J., Nilges, M., and Pastore, A. Structure of the dsRNA binding domain of E. coli RNase III. *EMBO J.* Vol.14(14):3572-3584

Kim, U., Wang, Y., Sanford ,T., Zeng, Y., and Nishikura, K. 1994. Molecular cloning of cDNAs for dsRNA adenosine deaminase, a candidate enzyme for nuclear RNA editing. *Proc Natl Acad Sci USA.* Vol.91:11457-11461

Kim, Y.G., Muralinath, M., Brandt, T., Pearcy, M., Hauns, K., Lowenhaupt, K., Jacobs, B.L., and Rich, A. 2003. A role for Z-DNA binding in vaccinia virus pathogenesis. *Proc Natl Acad Sci USA.* Vol.100(12):6974-6979

Kiebler, M.A., Hemraj, I., Verkade, P., Kohrmann, M., Fortes, P., Marion, R.M., Ortin, J., and Dotti, C.G. 1999. The mammalian staufen protein localizes to the somatodendritic domain of cultured hippocampal neurons: implications for its involvement in mRNA transport. *J Neurosci.* Vol.19(1):288-297

Knight, S.W., and Bass, B.L. 2002. The role of RNA editing by ADARs in RNAi. *Mol Cell.* Vol.10(4):809-817

Kobe, B. 1999. Autoinhibition by an internal nuclear localization signal revealed by the crystal structure of mammalian importin alpha. *Nat struct biol.* Vol.6:388-397

Komeili, A., and O'Shea, E.K. 2001. New perspectives on nuclear transport. *Annu Rev Genet.* Vol.35:341-364

Krovat, B.C., and Jantsch, M.F. 1996. Comparative mutational analysis of dsRBDs of Xenopus laevis RNA-binding protein A. *J Biol Chem.* Vol.271(45):28112-28119

Kumar, M., and Carmichael, G.G. 1997. Nuclear antisense RNA induces extensive adenosine modifications and nuclear retention of target transcripts. *Proc Natl Acad Sci USA.* Vol.94(8):3542-3547

Kuo, M.Y.P., Chao, M., and Taylor, J. 1989. Initiation of replication of the human hepatitis delta virus genome from cloned DNA: role of delta antigen. *J Virol.* Vol.63:1945-1950

Kutay, U., Bischoff, F.R., Kostka, S., Kraft, R., and Görlich, D. 1997. Export of importin alpha from the nucleus is mediated by a specific nuclear transport factor. *Cell.* Vol.90:1061–1071

Kutay, U., and Güttinger, S. 2005. Leucine-rich nuclear-export signals: born to be weak. *Trends Cell Biol.* Vol.15:121-124

Lai, F., Drakas, R., and Nishikura, K. 1995. Mutagenic analysis of double-stranded RNA adenosine deaminase, a candidate enzyme for RNA editing of glutamate-gated ion channel transcripts. *J Biol Chem.* Vol.270(29):17098-17105

Lamontagne, B., Tremblay, A., and Abou Elela, S. 2000. The N-terminal domain that distinguishes yeast from bacterial RNase III contains a dimerization signal required for efficient double-stranded RNA cleavage. *Mol Cell Biol.* Vol.20(4):1104-1115

Latimer, M., Ernst, M.K., Dunn, L.L., Drutskaya, M., and Rice, N.R. 1998. The N-terminal domain of IκBα masks the nuclear localisation signal of p50 and c-Rel homodimers. *Mol Cell Biol.* Vol.18(5):2640-2649

Lee, Y., Ahn, C., Han, J., Choi, H., Kim, J., Yim, J., Lee, J., Provost, P., Radmark, O., Kim, S., and Kim, V.N. 2003. The nuclear RNase III Drosha initiates microRNA processing. *Nature*. Vol.425(6956):415-419

Lehmann, K.A., and Bass, B.L. 2000. Double-stranded RNA adenosine deaminases ADAR1 and ADAR2 have overlapping specificities. *Biochemistry*. Vol.39(42):12875-12884

Lei, E.P., and Silver, P.A. 2002. Protein and RNA export from the nucleus. *Dev Cell*. Vol.2(3):261-272

Lellek, H., Kirsten, R., Diehl, I., Apostel, F., Buck, F., and Greeve, J. 2000. Purification and molecular cloning of a novel essential component of the apolipoprotein B mRNA editing enzyme-complex. *J Biol Chem*. Vol.275(26):19848-19856

Leulliot, N., Quevillon-Cheruel, S., Graille, M., van Tilbeurgh, H., Leeper, T.C., Godin, K.S., Edwards, T.E., Sigurdsson, S.T.L., Rozenkrants, N., Nagel, R.J., Ares Jr., M., and Varani, G. 2004. A new α-helical extension promotes RNA binding by the dsRBD of Rnt1p RNAseIII. *EMBO J*. Vol.23:2468-2477

Levanon, E.Y., Eisenberg, E., Yelin, R., Nemzer, S., Hallegger, M., Shemesh, R., Fligelman, Z.Y., Shoshan, A., Pollock, S.R., Sztybel, D., Olshansky, M., Rechavi, G., and Jantsch, M.F. 2004. Systematic identification of abundant A-to-I editing sites in the human transcriptome. *Nat Biotechnol*. Vol.22(8):1001-1005

Levanon, E.Y., Hallegger, M., Kinar, Y., Shemesh, R., Djinovic-Carugo, K., Rechavi, G., Jantsch, M.F., and Eisenberg, E. 2005. Evolutionarily conserved human targets of adenosine to inosine RNA editing. *Nucleic Acids Res*. Vol.33(4):1162-1168

Lindsay, M.E., Plafker, K., Smith, A.E., Clurman, B.E., and Macara, I.G. 2002. Npap60/Nup50 is a tri-stable switch that stimulates importin-alpha:beta-mediated nuclear protein import. *Cell*. Vol.110:349-360

Liu, Y., and Samuel, C.E. 1996. Mechanism of interferon action: functionally distinct RNA-binding and catalytic domains in the interferon-inducible, double-stranded RNA-specific adenosine deaminase. *J Virol*. Vol.70(3):1961-1968

Liu, Y., George, C.X., Patterson, J.B., and Samuel, C.E. 1997. Functionally distinct double-stranded RNA-binding domains associated with alternative splice site variants of the interferon-inducible double-stranded RNA-specific adenosine deaminase. *J Biol Chem.* Vol.272(7):4419-4428

Liu, Y., Emeson, R.B., and Samuel, C.E. 1999. Serotonin-2C receptor pre-mRNA editing in rat brain and in vitro by splice site variants of the interferon-inducible double-stranded RNA-specific adenosine deaminase ADAR1. *J Biol Chem.* Vol.274(26):18351-18358

Liu, Q., Liu, W., Jiang, L., Sun, M., Ao, Y., Zhao, X., Song, Y., Luo, Y., Lo, W.H., and Zhang, X. 2004. Novel mutations of the RNA-specific adenosine deaminase gene (DSRAD) in Chinese families with dyschromatosis symmetrica hereditaria. *J Invest Dermatol.* Vol.122(4):896-899

Lomeli, H., Mosbacher, J., Melcher, T., Höger, T., Geiger, J.R.P., Kuner, T., Monyer, H., Higuchi, M., Bach, A., and Seeburg, P.H. 1994. Control of kinetic properties of AMPA receptor channels by nuclear RNA editing. *Science.* Vol.266:1709-1712

Luciano, D.J., Mirsky, H., Vendetti, N.J., and Maas, S. 2004. RNA editing of a miRNA precursor. *RNA.* Vol.10(8):1174-1177

Lund, E., Güttinger, S., Calado, A., Dahlberg, J.E., and Kutay, U. 2003. Nuclear export of microRNA precursors. *Science.* Vol.303(5654):95-98

Lyman, S.K., and Gerace, L. 2001. Nuclear pore complexes: dynamics in unexpected places. *J Cell Biol.* Vol.154(1):17-20

Maas, S., Gerber, A.P., and Rich, A. 1999. Identification and characterization of a human tRNA-specific adenosine deaminase related to the ADAR family of pre-mRNA editing enzymes. *Proc Natl Acad Sci USA.* Vol.96(16):8895-8900

Maas, S., Kim, Y.G., and Rich, A. 2000. Sequence, genomic organization and functional expression of the murine tRNA-specific adenosine deaminase ADAT1. *Gene.* Vol.43(1-2):59-66

Maas, S., and Rich, A. 2000. Changing genetic information through RNA editing. *Bioessays.* Vol.22(9):790-802

Maas, S., Patt, S., Schrey, M., and Rich, A. 2001. Underediting of glutamate receptor GluR-B mRNA in malignant gliomas. *Proc Natl Acad Sci USA.* Vol.98(25):14687-14692

Macchi, P., Brownawell, A., Grunewald, B., DesGroseillers, L., Macara, I.G., and Kiebler, M.A., 2004. The Brain-specific Double-stranded RNA-binding Protein Staufen2. *J Biol Chem.* Vol.279:31440–31444

Martel, C., Macchi, P., Furic, L., Kiebler, M.A., and Desgroseillers, L. 2005. Staufen1 is imported into the nucleolus via a bipartite nuclear localization signal and several modulatory determinants. *Biochem J.* Immediate Publication on 15 Sep 2005.

Mattaj, I.W., and Englmeier, L. 1998. Nucleocytoplasmic transport: the soluble phase. *Annu Rev Biochem.* Vol.67:265-306

Mehta, A., Kinter, M.T., Sherman, N.E., and Driscoll, D.M. 2000. Molecular cloning of apobec-1 complementation factor, a novel RNA-binding protein involved in the editing of apolipoprotein B mRNA. *Mol Cell Biol.* Vol.20(5):1846-1854

Melcher, T., Maas, S., Herbert, A., Sprengel, R., Higuchi, M., and Seeburg, P.H. 1996. RED2, a brain specific member of the RNA-specific adenosine deaminase family. *J Biol Chem.* Vol.271:31795-31798

Michael, W.M. 2000. Nucleocytoplasmic shuttling signals: two for the price of one. *Trends Cell Biol.* Vol.10:46-50

Micklem, D.R., Adams, J., Grunert, S., and St Johnston, D. 2000. Distinct roles of two conserved Staufen domains in oskar mRNA localization and translation. *EMBO J.* Vol.19(6):1366-1377

Mittaz, L., Antonarakis, S.E., Higuchi, M., and Scott, H.S. 1997. Localization of a novel human RNA-editing deaminase (hRED2 or ADARB2) to chromosome 10p15. *Hum Genet.* Vol.100(3-4):398-400

Morse, D.P., Aruscavage, P.J., and Bass, B.L. 2002. RNA hairpins in noncoding regions of human brain and Caenorhabditis elegans mRNA are edited by adenosine deaminases that act on RNA.

Mühlhäusser, P., Müller, E.-C., Otto, A., and Kutay, U. 2001. Multiple pathways contribute to nuclear import of core histones. *EMBO Rep.* Vol.2(8):690-696

Muramatsu, M., Kinoshita, K., Fagarasan, S., Yamada, S., Shinkai, Y., and Honjo, T. 2000. Class switch recombination and hypermutation require activation-induced cytidine deaminase (AID), a potential RNA editing enzyme. *Cell.* Vol.102(5):553-63

Nachury, M.V., Maresca, T.J., Salmon, W.C., Waterman-Storer, C.M., Heald, R., and Weis, K. 2001. Importin beta is a mitotic target of the small GTPase Ran in spindle assembly. *Cell.* Vol.104:95-106

Nakielny, S., and Dreyfuss, G. 1999. Transport of proteins and RNAs in and out of the nucleus. *Cell.* Vol.99:677-690

Nie, Y., Zhoa, Q., Su, Y., and Yang, J-H. 2004. Subcellular distribution of ADAR1 isoforms is synergistically determined by three nuclear discrimination signals and a regulatory motif. *J Biol Chem.* Vol.279:13249-13255

Nishikura, K., Yoo, C., Kim, U., Murray, G.M., Estes, P.A., Cash, F.E., and Liebhaber, F.A. 1991. Substrate specificity of the dsRNA unwinding/modifying activity. *EMBO J.* Vol.10:3523-3532

Niswender, C.M., Sanders-Bush, E., and Emeson, R.B. 1998. Identification and characterization of RNA editing events within the 5-HT2C receptor. *Ann NY Acad Sci.* Vol.861:38-48

Niswender, C.M., Herrick-Davis, K., Dilley, G.E., Meltzer, H.Y., Overholser, J.C., Stockmaier, C.A., Emeson, R.B., and Sanders-Bush, E. 2001. RNA-editing of the human serotonin 5-HT$_{2C}$ receptor: alteration in suicide and implications for serotonergic pharmacotheray. *Neuropsychopharm.* Vol.24:478-491

O'Connell, M.A., and Keller, W. 1994. Purification and properties of double-stranded RNA-specific adenosine deaminase from calf thymus. *Proc Natl Acad Sci USA.* Vol.91(22):10596-10600

O'Connell, M.A., Gerber ,A., and Keller, W. 1997. Purification of human double-stranded RNA-specific editase1 (hRed1), involved in editing of brain glutamate receptor B pre-mRNA. *J Biol Chem.* Vol.272:473-478

Ohtsubo, M., Okazaki, H., and Nishimoto, T. 1989. The RCC1 protein, a regulator for the onset of chromosome condensation locates in the nucleus and binds to DNA. *J Cell Biol.* Vol.109:1389-1397

Orphanides, G. and Reinberg, D. 2002. A unified theory of gene expression. *Cell.* Vol.108(4):439-451

Palladino, M.J., Keegan, L.P., O'Connell, M.A., and Reenan, R.A. 2000. dADAR, a Drosophila double-stranded RNA-specific adenosine deaminase is highly developmentally regulated and is itself a target for RNA editing. *RNA.* Vol.6(7):1004-1018

Pante, N., and Kann, M. 2002. Nuclear pore complex is able to transport macromolecules with diameters of about 39 nm. *Mol Biol Cell.* Vol.13(2):425-434

Patel, R.C., and Sen, G.C. 1998. Requirement of PKR dimerization mediated by specific hydrophobic residues for its activation by double-stranded RNA and its antigrowth effects in yeast. *Mol Cell Biol.* Vol.18(12):7009-7019

Park, M.Y., Wu, G., Gonzalez-Sulser, A., Vaucheret, H., and Poethig, R.S. 2005. Nuclear processing and export of microRNAs in Arabidopsis. *Proc Natl Acad Sci USA.* Vol.102(10):3691-3696

Patterson, J.B., and Samuel, C.E. 1995. Expression and regulation by interferon of a double-stranded-RNA-specific adenosine deaminase from human cells: Evidence for two forms of the deaminase. *Mol Cell Biol.* Vol.15:5376-5388

Patton, D., Silva, T., and Bezanilla, F. 1997. RNA editing generates a diverse array of transcripts encoding squid Kv2 K+ channels with altered functional properties. *Neuron* Vol.19:711-722.

Paul, M., and Bass, B.L. 1998. Inosine exists in mRNA at tissue-specific levels and is most abundant in brain mRNA. *EMBO J.* Vol.17:1120-1127

Penalva, L.O., and Sanchez, L. 2003. RNA binding protein sex-lethal (Sxl) and control of Drosophila sex determination and dosage compensation. *Microbiol Mol Biol Rev.* Vol.67(3):343-359

Pinol-Roma, S., and Dreyfuss, G. 1999. Shuttling of pre-mRNA binding proteins between nucleus and cytoplasm. *Nature.* Vol.355:730-732

Pollard, V.W., Michael, W.M., Nakielny, S., Siomi, M.C., Wang, F., and Dreyfuss, G. 1996. A novel receptor-mediated nuclear protein import pathway. *Cell.* Vol.86:985–994

Polson, A.G., Crain, P.F., Pomerantz, S.C., McCloskey, J.A., and Bass, B.L. 1991. The mechanism of adenosine to inosine conversion by the double-stranded RNA unwinding/modifying activity: a high-performance liquid chromatography-mass spectrometry analysis. *Biochemistry.* Vol.30:11507-11514

Polson, A.G., and Bass, B.L. 1994. Preferential selection of adenosines for modification by double-stranded RNA adenosine deaminase. *EMBO J.* Vol.13:5701-5711

Polson, A.G., Bass, B.L., and Casey, J.L. 1996. RNA editing of hepatitis delta virus antigenome by dsRNA-adenosine deaminase. *Nature.* Vol. 380(6573):454-456. Erratum 1996 *Nature.* Vol.381(6580):346.

Poulsen, H., Nilson, J., Damgaard, C.K., Egebjerg, J., and Kjems, J. 2001. Crm1 mediates the export of ADAR1 through a nuclear export signal within the Z-DNA binding domain. *Mol Cell Biol.* Vol.21(22):7862-7871

Powell, L.M., Wallis, S.C., Pease, R.J., Edwards, Y.H., Knott, T.J., and Scott, J. 1987. A novel form of tissue-specific RNA processing produces apolipoprotein-B48 in intestine. *Cell.* Vol.50(6):831-840

Rebagliati, M.R., and Melton, D.A. 1987. Antisense RNA injections in fertilized frog eggs reveal an RNA duplex unwinding activity. *Cell.* Vol.48:599-605

Rebane, A., Aab, A., and Steitz, J.A. 2004. Transportins 1 and 2 are redundant nuclear import factors for hnRNP A1 and HuR. *RNA*. Vol.10(4):590-599

Rexach, M., and Blobel, G. 1995. Protein import into nuclei: association and dissociation reactions involving transport substrate, transport factors, and nucleoporins. *Cell.* Vol.83:683–692

Ribbeck, K., Lipowsky, G., Kent, H.M., Stewart, M., and Görlich, D. 1998. NTF2 mediates nuclear import of Ran. *EMBO J.* Vol.17:6587-6598

Ribbeck, K., and Görlich, D. 2001. Kinetic analysis of translocation through nuclear pore complexes. *EMBO J.* Vol.20(6):1320-1230

Robbins, J., Dilworth, S.M., Laskey, R.A., and Dingwall, C. 1991. Two interdependent basic domains in nucleoplasmin nuclear targeting sequence: identification of a class of bipartite nuclear targeting sequence. *Cell.* Vol.64(3):615-623

Romano, P.R., Zhang, F., Tan, S.L., Garcia-Barrio, M.T., Katze, M.G., Dever, T.E., and Hinnebusch, A.G. 1998. Inhibition of double-stranded RNA-dependent protein kinase PKR by vaccinia virus E3: role of complex formation and the E3 N-terminal domain. *Mol Cell Biol.* Vol.18(12):7304-73016

Rout, M.P., Aitchison, J.D., Suprapto, A., Hjertaas, K., Zhao, Y., and Chait, B.T. 2000. The yeast nuclear pore complex: composition, architecture, and transport mechanism. *J Cell Biol.* Vol.148:635-651

Rueter, S.M., Dawson, T.R., and Emeson, R.B. 1999. Regulation of alternative splicing by RNA editing. *Nature.* Vol.399:75-80

Ryter, J.M., and Schultz, S.C. 1998. Molecular basis of double-stranded RNA-protein interactions: structure of a dsRNA-binding domain complexed with dsRNA. *EMBO J.* Vol.17(24):7505-7513

Sallacz, N.B., and Jantsch, M.F. 2005. Chromosomal storage of the RNA-binding enzyme ADAR1 in *Xenopus* oocytes. *Mol Biol Cell.* Vol.16:3377-3386

Sansam, C.L., Wells, K.S., and Emeson, R.B. 2003. Modulation of RNA editing by functional nucleolar sequestration of ADAR2. *Proc Natl Acad Sci USA.* Vol.100:14018-14023

Saunders, L.R., and Barber, G.N. 2003. The dsRNA binding protein family: critical roles, diverse cellular functions. *FASEB*. Vol.17:961-983

Scadden, A.D. and Smith, C.W. 1997. A ribonuclease specific for inosine-containing RNA: a potential role in antiviral defence? *EMBO J*. Vol.16(8):2140-2149

Scadden, A.D. and Smith, C.W. 2001. Specific cleavage of hyper-edited dsRNAs. *EMBO J*. Vol.20(15):4243-4252

Schade, M., Behlke, J., Lowenhaupt, K., Herbert, A., Rich, A., and Oschkinat, H. 1999a. A 6 bp Z-DNA hairpin binds two Z alpha domains from the human RNA editing enzyme ADAR1. *FEBS Lett*. Vol.458(1):27-31

Schade, M., Turner, C.J., Kuhne, R., Schmieder, P., Lowenhaupt, K., Herbert, A., Rich, A., and Oschkinat, H. 1999b. The solution structure of the Zalpha domain of the human RNA editing enzyme ADAR1 reveals a prepositioned binding surface for Z-DNA. *Proc Natl Acad Sci USA*. Vol.96(22):12465-12470

Schwartz, T., Rould, M.A., Lowenhaupt, K., Herbert, A., and Rich, A. 1999. Crystal structure of the Z domain of the human editing enzyme ADAR1 bound to left-handed Z-DNA. *Science*. Vol.284:1841-1845

Shah, S., Tugendreich, S., and Forbes, D. 1998. Major binding sites for the nuclear import receptor are the internal nucleoporin Nup153 and the adjacent nuclear filament protein Tpr. *J Cell Biol*. Vol.141:31-40

Shyu, A.B., and Wilkinson, M.F. 2000. The double lives of shuttling mRNA binding proteins. *Cell*. Vol.102:135-138

Siomi, M..C., Eder, P.S., Kataoka, N., Wan, L., Liu, Q., and Dreyfuss, G. 1997. Transportin-mediated nuclear import of heterogeneous nuclear RNP proteins. *J Cell Biol*. Vol.138:1181–1192

Smith, A., Brownawell, A., and Macara, I.G. 1998. Nuclear import of Ran is mediated by the transport factor NTF2. *Curr Biol.* Vol.8:1403:1406

Smith, A.E., Slepchenko, B.M., Schaff, J.C., Loew, L.M., and Macara, I.G. 2002. Systems analysis of Ran transport. *Science.* Vol.295(5554):488-491

Sommer, B., Kohler, M., Sprengel, R., and Seeburg, P.H. 1991. RNA editing in brain controls a determinant of ion flow in glutamate-gated channels. *Cell.* Vol.67(1):11-19

Stefl, R., Skrisovska, L., Xu, M., Emeson, R.B., and Allain, F.H. 2005. Resonance assignments of the double-stranded RNA-binding domains of adenosine deaminase acting on RNA 2 (ADAR2). *J Biomol NMR.* Vol.31(1):71-72

Stephens, O.M., Yi-Brunozzi, H.Y., and Beal, P.A. 2000. Analysis of the RNA-editing reaction of ADAR2 with structural and fluorescent analogues of the GluRB R/G editing site. *Biochem.* Vol.39:12243-12251

Stephens, O.M., Haudenschild, B.L., and Beal, P.A. 2004. The binding selectivity of ADAR2's dsRBMs contributes to RNA-editing selectivity. *Chem Biol.* Vol.11(9):1239-1250

St Johnston, D., Brown, N.H., Gall, J.G., and Jantsch, M. 1992. A conserved double-stranded RNA-binding domain. *Proc Natl Acad Sci USA.* Vol.89:10979-10983

Stoffler, D., Feja, B., Fahrenkrog, B., Walz, J., Typke, D., and Aebi, U. 2003. Cryo-electron tomography provides novel insights into nuclear pore architecture: implications for nucleocytoplasmic transport. *J Mol Biol.* Vol.328:119-130

Strehblow, A., M. Hallegger, and M.F. Jantsch. 2002. Nucleocytoplasmic distribution of human RNA-editing enzyme ADAR1 Is modulated by double-stranded RNA-binding domains, a leucine-rich export signal, and a putative dimerization domain. *Mol Biol Cell.* 13:3822-35

Ström, A.C., and Weis, K. 2001. Importin-beta-like nuclear transport receptors. *Genome Biol.* Vol.2(6):3008.1-3008.9

Taylor, J.M. 1999. Replication of human hepatitis delta virus: influence of studies on subviral plant pathogens. *Adv Virus Res.* Vol.54:45-60

Taylor, D.R., Puig, M., Darnell, M.E., Mihalik, K., and Feinstone, S.M. 2005. New antiviral pathway that mediates hepatitis C virus replicon interferon sensitivity through ADAR1. *J Virol.* Vol.79(10):6291-6298

Teng, B., Burant, C.F., and Davidson, N.O. 1993. Molecular cloning of an apolipoprotein B messenger RNA editing protein. *Science.* Vol.60(5115):1816-1819

Tian, B., Bevilacqua, P.C., Diegelman-Parente, A., and Mathews, M.B. 2004. The double-stranded-RNA-binding motif: interference and much more. *Nat Rev Mol Cell Biol.* Vol.5(12):1013-1023

Tonkin, L.A., Saccomanno, L., Morse, D.P., Brodigan, T., Krause, M., and Bass, B.L. 2002. RNA editing by ADARs is important for normal behavior in Caenorhabditis elegans. *EMBO J.* Vol.21(22):6025-6035

Tonkin, L.A., and Bass, B.L. 2003. Mutations in RNAi rescue aberrant chemotaxis of ADAR mutants. *Science.* Vol.302(5651):1725

Truant, R., and Cullen, B.R. 1999. The argenine-rich domains present in Human Immunodeficiency Virus type 1 Tat and Rev function as direct Importin β-dependent nuclear localisation signals. *Mol Cell Biol.* Vol.19(2):1210-1217

Wagner, R.W., and Nishikura, K. 1988. Cell cycle expression of RNA duplex unwindase activity in mammalian cells. *Mol Cell Biol.* Vol.8:770-777

Wagner, R.W., Smith, J.E., Cooperman, B.S., and Nishikura, K. 1989. A double-stranded RNA unwinding activity introduces structural alterations by means of adenosine to inosine conversions in mammalian cells and *Xenopus* eggs. *Proc Natl Acad Sci USA.* Vol.86:2647-2651

Wang, Q., Khillan, J., Gadue, P., and Nishikura, K. 2000. Requirement of the RNA editing deaminase ADAR1 gene for embryonic erythropoiesis. *Science.* Vol.290:1765-1768

Wang, Q., Miyakoda, M., Yang, W., Khillan, J., Stachura, D.L., Weiss, M.J., and Nishikura, K. 2004. Stress-induced apoptosis associated with null mutation of ADAR1 RNA editing deaminase gene. *J Biol Chem.* Vol. 279(6):4952-4961

Wang, Q., and Carmichael, G.G. 2004. Effects of length and location on the cellular response to double-stranded RNA. *Microbiol Mol Biol Rev.* Vol.68(3):432-52

Wang, Q., Zhang, Z., Blackwell, K., and Carmichael, G.G. 2005. Vigilins bind to promiscuously A-to-I-edited RNAs and are involved in the formation of heterochromatin. *Curr Biol.* Vol.15(4):384-391

Watson, M.L. 1959. Further observations on the nuclear envelope of the animal cell. *J Biophys Biochem Cytol.* Vol.6:147-156

Weighardt, F., Biamonti, G., and Riva, S. 1995. Nucleo-cytoplasmic distribution of human hnRNP proteins: a search for the targeting domains in hnRNP A1. *J Cell Sci.* Vol.108(2):545-555

Wen, W., Meinkoth, J.L., Tsien, R.Y., and Taylor, S.S. 1995. Identification of a signal for rapid export of proteins from the nucleus. *Cell.* Vol.82(3):463-473

Wolf, J., Gerber, A.P., and Keller, W. 2002. tadA, an essential tRNA-specific adenosine deaminase from Escherichia coli. *EMBO J.* Vol.21(14):3841-3851

Wong, T.C., Ayata, M., Hirano, A., Yoshikawa, Y., Tsuruoka, H., and Yamanouchi, K. 1989. Generalized and localized biased hypermutation affecting the matrix gene of a measles virus strain that causes subacutesclerosing panencephalitis. *J Virol.* Vol.63(12):5464-5468

Wong, S.K., Sato, S., and Lazinski, D.W. 2001. Substrate recognition by ADAR1 and ADAR2. *RNA.* Vol.7(6):846-858

Wong, S.K., S. Sato, and D.W. Lazinski. 2003. Elevated activity of the large form of ADAR1 in vivo: Very efficient RNA editing occurs in the cytoplasm. *RNA.* Vol.9:586-598

Wu, S., and Kaufman, R.J. 1997. A model for the double-stranded RNA (dsRNA)-dependent dimerization and activation of the dsRNA-activated protein kinase PKR. *J Biol Chem.*

Vol.272(2):1291-1296

Yi, R., Qin, Y., Macara, I.G., and Cullen, B.R. 2003. Exportin-5 mediates the nuclear export of pre-miRNAs and short hairpin RNAs. *Genes Dev.* Vol.17:3011-3016

Yi, R., Doehle, B.P., Qin, Y., Macara, I.G., and Cullen, B.R. 2005. Overexpression of exportin 5 enhances RNA interference mediated by short hairpin RNAs and microRNAs. *RNA.* Vol.11(2):220-226

Yi-Brunozzi, H.Y., Stephens, O.M., and Beal, P.A. 2001. Conformational changes that occur during an RNA editing adenosine deamination reaction. *J Biol Chem.* Vol.276:37827-37833

Yokoyama, N., Hayashi, N., Seki, T., Panté, N., Ohba, T., Nishii, K., Kuma, K., Hayashida, T., Miyata, T., Aebi, U., Fukui, M., and Nishimoto, T. 1995. A giant nucleopore protein that binds Ran/TC4. *Nature.* Vol.376:184-188

Yoshida, K., and Blobel, G. 2001. The karyopherin Kap142p/Msn5p mediates nuclear import and nuclear export of different cargo proteins. *J Cell Biol.* Vol.152(4):729-740

Zhang, C., and Clarke, P.R. 2000. Chromatin-independent nuclear envelope assembly induced by Ran GTPase in Xenopus egg extracts. *Science.* Vol.288:1429-1432

Zhang, Z., and Carmichael, G.G. 2001. The fate of dsRNA in the nucleus: a p54(nrb)-containing complex mediates the nuclear retention of promiscuously A-to-I edited RNAs. *Cell.* Vol.106(4):465-475

Many thanks to ....

First, I want to thank my supervisor, Michael Jantsch, for his ideas and support during the last five years.

Thanks to all members of the Jantsch lab for the great atmosphere and friendship, especially to Jutta and Sebastian who also contributed to this work.

Many thanks also to my family and friends for their support and keeping my motivation high.

This work is dedicated to my wife Tanja.
If not for you, my sky would fall. I love you!

My thesis was supported by the Austrian Academy of Sciences.

VDM Verlagsservicegesellschaft mbH

Die VDM Verlagsservicegesellschaft sucht für wissenschaftliche Verlage abgeschlossene und herausragende

Dissertationen, Habilitationen, Diplomarbeiten, Master Theses, Magisterarbeiten usw.

für die kostenlose Publikation als Fachbuch.

Sie verfügen über eine Arbeit, die hohen inhaltlichen und formalen Ansprüchen genügt, und haben Interesse an einer honorarvergüteten Publikation?

Dann senden Sie bitte erste Informationen über sich und Ihre Arbeit per Email an *info@vdm-vsg.de*.

Sie erhalten kurzfristig unser Feedback!

VDM Verlagsservicegesellschaft mbH
Dudweiler Landstr. 99   Telefon +49 681 3720 174
D - 66123 Saarbrücken   Fax   +49 681 3720 1749
www.vdm-vsg.de

Die VDM Verlagsservicegesellschaft mbH vertritt

Printed by Books on Demand GmbH, Norderstedt / Germany